Construction Depth

PE

Civil Engineering Exams

Two Full Exams with Solutions

Md Rashad Islam, Ph.D., P.E.

Dr. Md Rashad Islam, P.E.
islamunm@gmail.com

Cover Photo by Armando Perez, Pueblo, Colorado

ISBN-13: 978-1-957186-04-7

First Edition, April 2022
Reprinting with minor revision, November 2022
Reprinting with minor revision, May 2023

Printed in the United States of America

Preface

This Construction Depth PE Civil Engineering Exam book contains 2 full sample exams (40 questions each) with detailed solutions for the Computer-Based Testing (CBT) of the PE Civil afternoon (depth) examination starting. *PE Civil Reference Handbook* and the other NCEES – recommended references have been primarily used to solve the problems. The location of the solutions' equations or theories in the *PE Civil Reference Handbook* and the references are also pointed out. The exam specification of Construction depth has been thoroughly checked to confirm that this book is most updated. The following topics are covered for the Construction depth exam (afternoon session):

9. Earthwork Construction and Layout (5–8 questions)
10. Estimating Quantities and Costs (5–8 questions)
11. Construction Operations and Methods (6–9 questions)
12. Scheduling (5–8 questions)
13. Material Quality Control and Production (5–8 questions)
14. Temporary Structures (6–9 questions)
15. Health and Safety (3–5 questions)

Topics 1. to 8. are covered in the PE Civil Engineering Breadth (morning) Exam.

The book has been thoroughly inspected with the help of professional editors to fix typos, editorial issues, and poor sentence structure. Despite this, if there is any issue, please excuse it and report it to the author. The author is also requesting the readers to report any concern about the book. Any suggestion to improve this book, or any issue reported, will be fixed in the next edition with proper acknowledgement. Thanking you.

Dr. Md Rashad Islam, P.E.
Associate Professor
Colorado State University – Pueblo
islamunm@gmail.com
https://www.csupueblo.edu/profile/md-islam/index.html

Resource Allowed

In addition to the ***PE Civil Reference Handbook***, the following codes and standards (most of them are available in Google) will be supplied to examinees on exam day as a searchable, electronic pdf file with linked chapters for easy navigation. Please go through all these resources so that you know which information is available and where it is available. You should be familiar with all these documents and be competent with the most common topics.

ACI 347R *Guide to Formwork for Concrete,* 2014, American Concrete Institute, Farmington Hills, MI, www.concrete.org.

ACI SP-4 *Formwork for Concrete,* 8/e, 2014, American Concrete Institute, Farmington Hills, MI, www.concrete.org.

AISC *Steel Construction Manual,* 14/e, 2011, American Institute of Steel Construction, Inc., Chicago, IL, www.aisc.org.

ASCE 37-14 *Design Loads on Structures During Construction*, 2/e, 2015, American Society of Civil Engineers, Reston, VA, www.asce.org.

CMWB *Standard Practice for Bracing Masonry Walls under Construction*, 2012, Council for Masonry Wall Bracing, Mason Contractors Association of America, Lombard, IL, www.masoncontractors.org.

MUTCD-Pt 6 *Manual on Uniform Traffic Control Devices for Streets and Highways — Part 6 Temporary Traffic Control,* 2009, including Revisions 1 and 2 dated May 2012, US Department of Transportation, Federal Highway Administration, Washington, DC, www.fhwa.dot.gov.

CFR TITLE 29 U.S. Department of Labor, Washington, D.C., July 2020.

Part 1903 Inspections, Citations, and Proposed Penalties

Part 1904 Recording and Reporting Occupational Injuries and Illnesses

Part 1926 Safety and Health Regulations for Construction

PCA EB001 *Design and Control of Concrete Mixtures*, 17/e, 2021, Portland Cement Association, Skokie, IL, www.cement.org.

Reference categories for Construction depth module:

- Construction surveying
- Construction estimating
- Construction planning and scheduling
- Construction equipment and methods
- Construction materials
- Construction design standards (see above)

Practice Exam 1

Number of Questions = 40
Time = 4.0 hours

Question 1.1

The following welding symbol is indicated on a construction drawing.

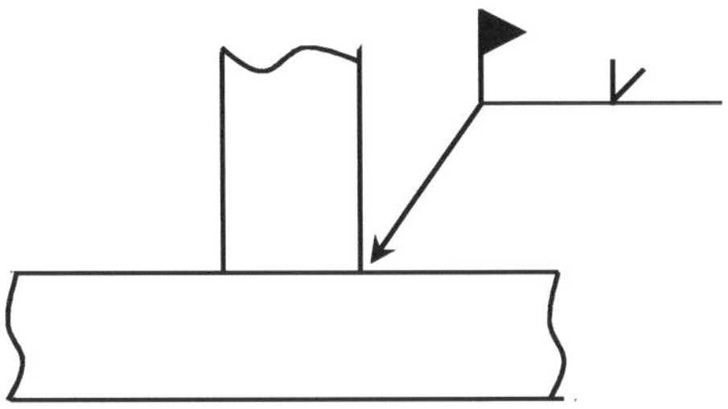

Select the correct weld deposit for this symbol.

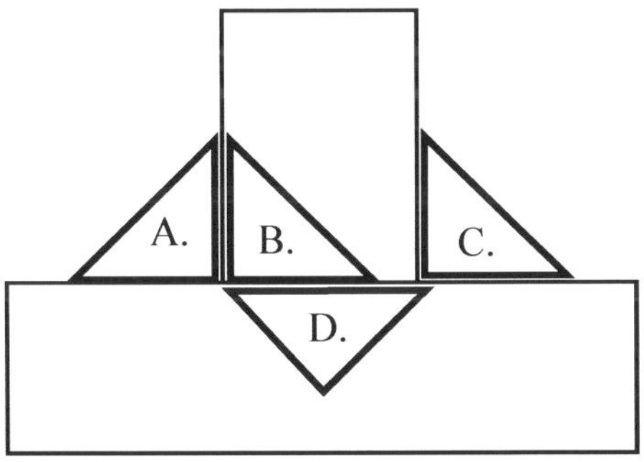

Question 1.2

You are working as a quality control representative for a contractor on a project. If you discover a potentially unsafe condition at the project site, the initial action you should take is to:

A. Stop the construction of the project
B. Report the condition to the contractor
C. Report the condition to the owner
D. Report the condition to OSHA

Question 1.3

The cut and fill data of a highway segment is listed below. The shrinkage of the soil is 25% from natural (bank) to compacted state.

Station	Cut (ft^2)	Fill (ft^2)
0	40	0
1	140	20
2	160	25
3	60	116

The first ordinate (circled) in the mass diagram shown below is most nearly:

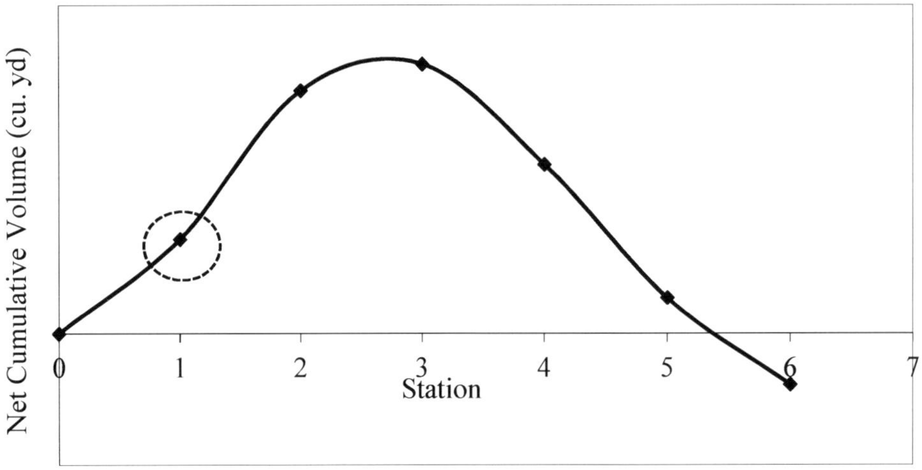

A. 49.3
B. 340
C. 333
D. 284

Question 1.4

A construction project has the following 4 activities with their durations and relationships shown. The minimum number of days required to finish the project is most nearly:

Activity	Duration	Immediately Preceding Activity	Relationship	Lag
A	3	-		
B	5	A	SS	3
C	2	A		
D	6	B	FF	
		C		

A. 8
B. 11
C. 9
D. 13

Question 1.5

A construction company has 800,000 employee-hour worked per year with the following injury records:

Minor injuries (first aid only) = 12
Medical-only injuries (no list time or light duty) = 5
Medical injuries in light duty restrictions = 4
Long-time injuries = 6

The OSHA incidence rate is most nearly:

A. 6.75
B. 3.5
C. 3.75
D. 4.0

Question 1.6

The plan view of a basement foundation is shown below. The given dimensions are on center-to-center. The wall is 1-ft thick and 10-ft high. The footing is 1.5-ft wide and 3-ft high. A wooden formwork is to be made for the walls for concrete casting. The required contact surface area (ft^2) of wall formwork is most nearly:

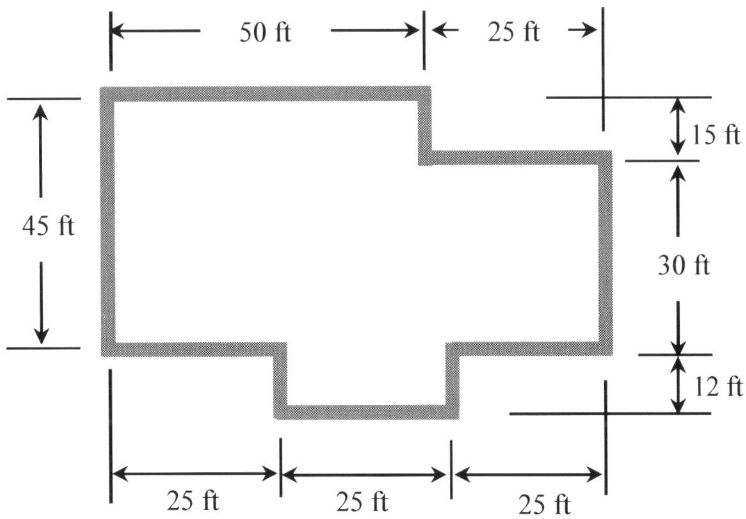

A. 5,280
B. 4,820
C. 5,560
D. 2,640

Question 1.7

The profile of a 4,000-mm clear-spanned reinforced-concrete beam is shown below. The columns supporting this beam is 230 mm square. The clear cover is 40 mm from all sides of members.

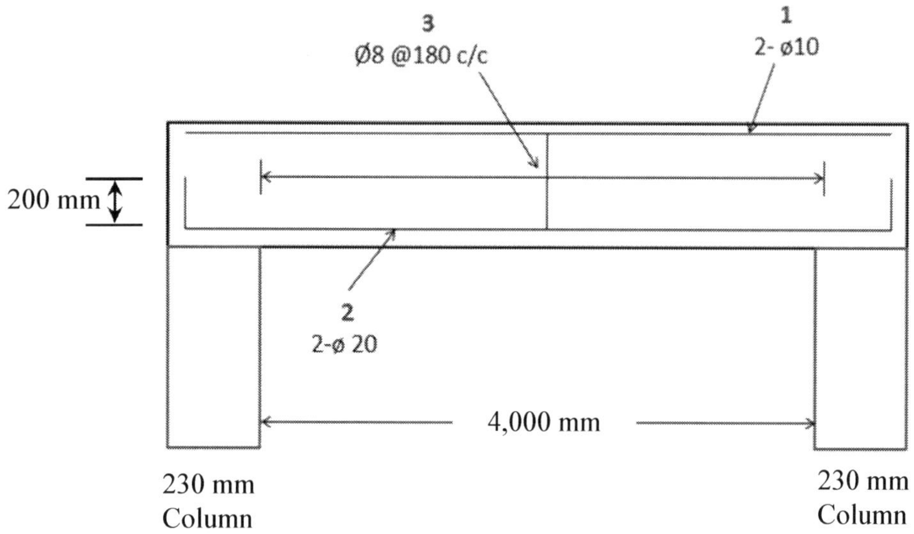

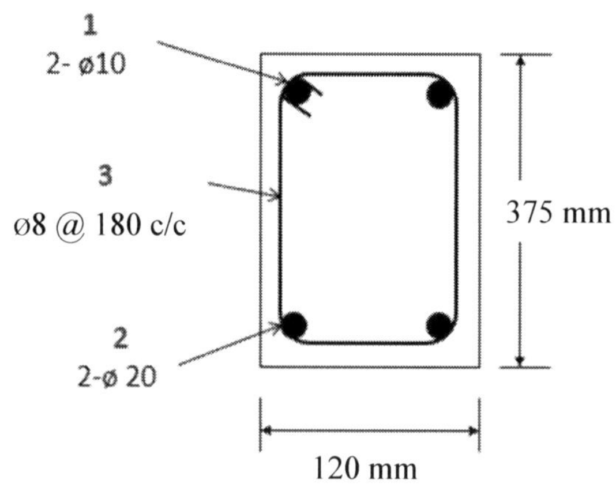

Ignore the length change due to the bending of bars. The cutting length (mm) of the top bar (Bar 1) and bottom bar (Bar 2), respectively are most nearly:

A. 4,300 and 4,780
B. 4,380 and 4,380
C. 4,150 and 4,550
D. 4,380 and 4,780

Question 1.8

A reinforced concrete retaining wall is required for a site. Reinforcement is placed on the tension side of a reinforced-concrete section. The reinforced concrete retaining wall section with the most efficient footing configuration and steel reinforcing is most nearly:

A.

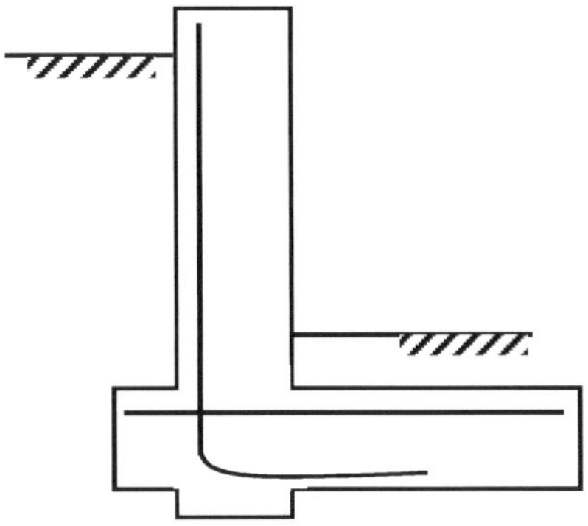

B.

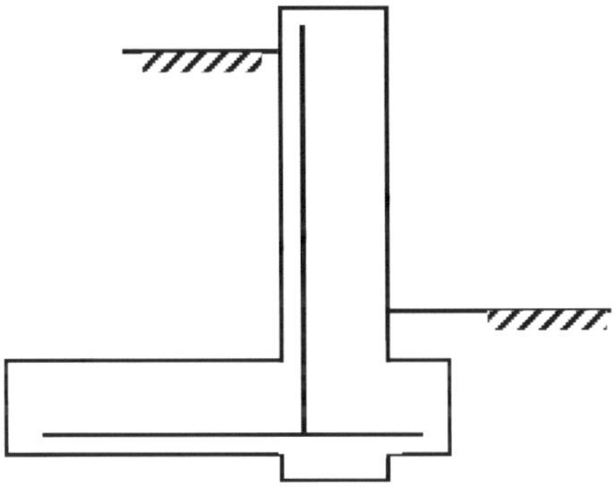

C.

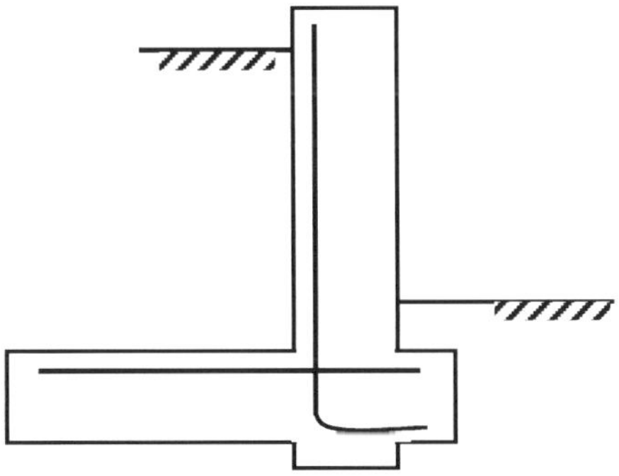

D.

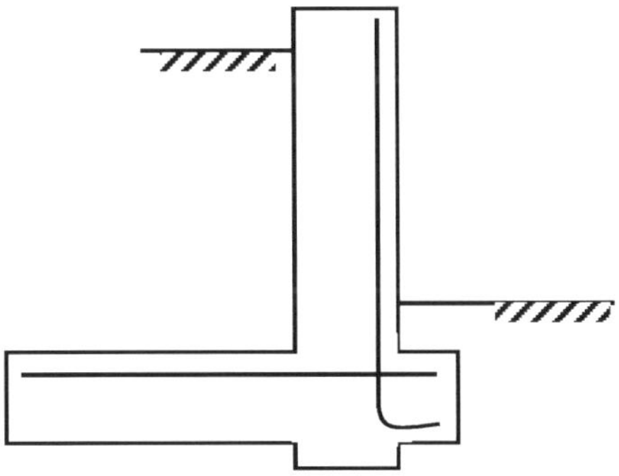

Question 1.9

A construction crew of six (6) members can form a 40 ft^2 of contact area of wooden formwork per crew-hour when working at less than 8 ft above the grade. At more than 8 ft above the grade, the production rate drops to 60%. The crew costs \$372 per hour. Wooden formworks for a freestanding rectangular tank of 90-ft long, 40-ft wide and 16-ft tall is to be made. The required total crew-hour is most nearly:

A. 364
B. 278
C. 104
D. 174

Question 1.10

Which of the following statements is NOT correct for the following visual grading stamp on a lumber piece?

A. Western Wood Products (WWP) as a company or supplier
B. Two (2) times Kiln-Dried (KD) and Heat Treated (HT)
C. Producing mill lumber is 358
D. Wood species is White Fir

Question 1.11

A vertical formwork is constructed for a reinforced column construction. After the addition of all admixtures, the slump was measured to be about 4.0 in. The internal vibration depth will be about 3.0 ft. The concrete was made with all-purpose cement (Type I) with normal-weight aggregate with no retarder. It is anticipated that the rate of placement of the fresh concrete is 3.0 ft/hour at the temperature of 70 °F .

The maximum lateral pressure (psf) on the concrete formwork is most nearly:

A. 536
B. 580
C. 635
D. 600

Question 1.12

Which of the following volumes is the largest?

A. Heaped volume
B. Water-line volume
C. Struck volume
D. Plate-line volume

Question 1.13

Based on the soil classification system found in the federal OSHA regulation Part 1926, Subpart P, the soil adjacent to an existing building may be classified as Type A, B or C. The excavation depth is 15 ft deep. The undisturbed perimeter strip (X) along the face of the building is most nearly:

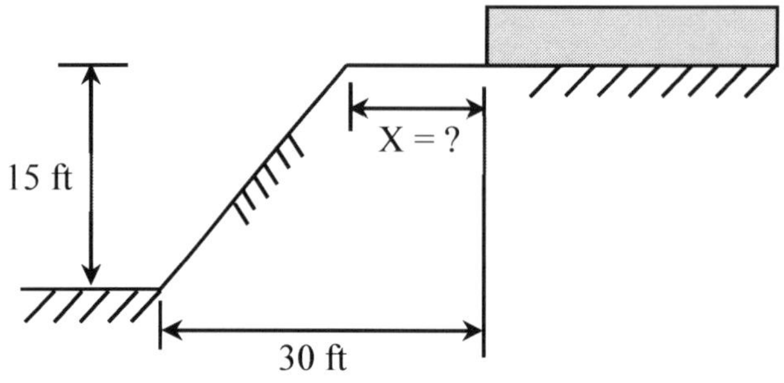

A. X = 13.75 ft for Type A soil
B. X = 10 ft for Type B soil
C. X = 7.5 ft for Type C soil
D. X = 18.75 ft for Type B soil

Question 1.14

It has been determined that a double-wythe clay masonry wall required 982 units without waste. All mortar joints are ½ in. thick. The actual unit dimensions is 3.75 x 2.25 x 8 in. Assuming a 25% waste factor, the quantity of mortar (ft^3) required for the wall is most nearly:

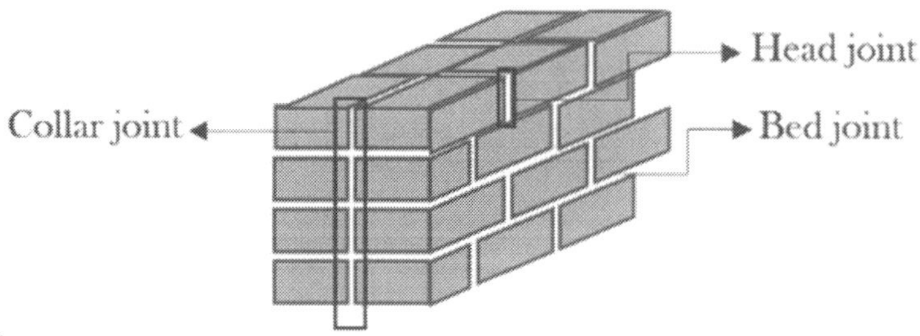

A. 12.5
B. 15.5
C. 18.5
D. 21.5

Question 1.15

A concrete mix will be used in foundation or basement walls extending above the grade that can experience freezing and thawing with frequent accumulation of snow and ice with deicing chemicals. The concrete will have a maximum *w/cm* ratio of 0.40 and minimum strength of 5,000 psi. The nominal maximum aggregate size is 1.0 in. Neglecting the tolerance, per PCA, the target air content (%) is most nearly:

A. 5.0
B. 5.5
C. 6.0
D. 7.0

Question 1.16

A downtown street requires a work-zone-lane shift to 12-ft as shown. The possible length (ft) for the shifting taper is 300 ft. The recommended work zone speed limit (mph) by MUTCD considering the safety is most nearly:

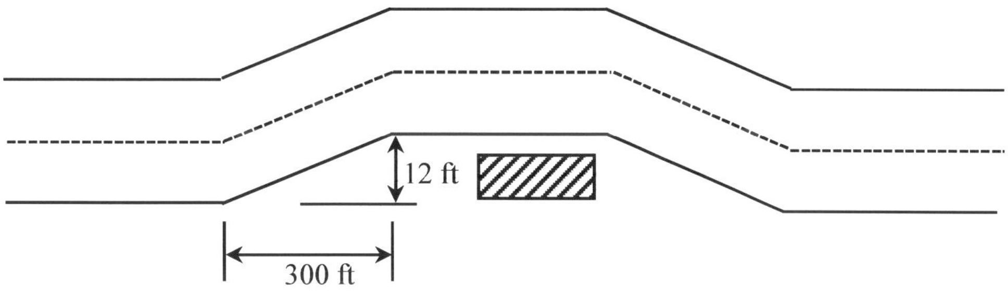

A. 20
B. 25
C. 35
D. 40

Question 1.17

Dummy activities in arrow diagrams are used for:

A. Straightening up the logic relationships
B. Distinguishing identities for activities that share the start and end nodes
C. Both A and B
D. None of the above

Question 1.18

A construction equipment is being supported by four cables as shown below. The plan view of the cabling system is rectangular, and the total load supported by the four-cable system is 30 kip. The slant length of each cable is 32 ft. All the cables carry equal load.

(a) Actual structure

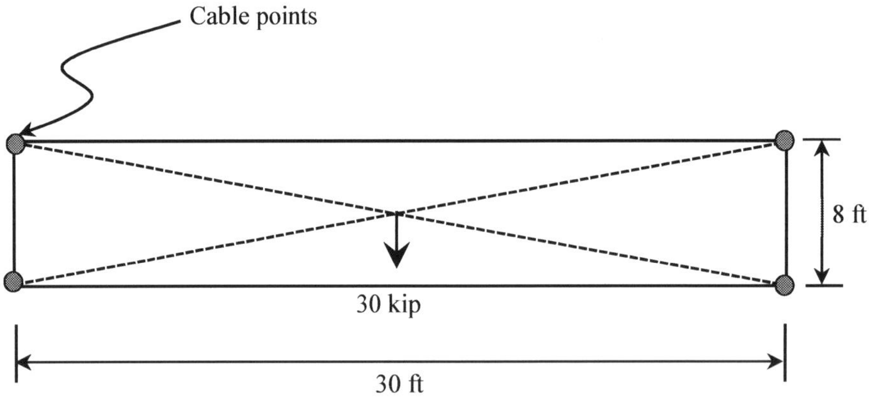

(b) Plan view of the cabling

The developed force (kip) in each of the four cables is most nearly:

A. 7.50
B. 8.57
C. 9.65
D. 6.56

Question 1.19

The following relationship for a precedence network means:

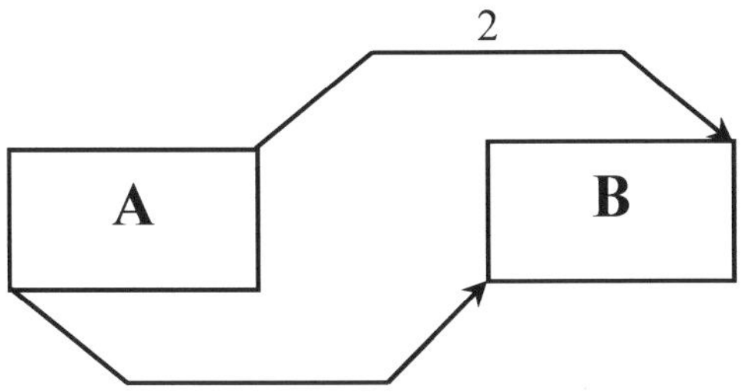

A. Activities A and B must start simultaneously, but B has to finish 2 days after the finish of A
B. Activity B can start after the start of A and but must finish at least 2 days before the finish of A
C. Activity B can start before the start of A but must finish at most 2 days after the finish of A
D. Activity B can start after the start of A and but must finish at least 2 days after the finish of A

Question 1.20

In the United States, a bag of Portland cement typically has a mass of _____kg and a volume of_______L.

A. 42 and 28 respectively
B. 94 and 1.0 respectively
C. 92.4 and 28 respectively
D. 42 and 1.0 respectively

Question 1.21

Assume a project engineer costs his employer $63 per hour ($50 wage + $13 benefits). Two projects are within 40 miles of each other. The engineer travels at an average speed of 40 miles per hour and costs the company $1 per mile to travel between the two projects. Assume that the following four statements are true:

- The engineer is needed exactly 3 hours per day in each project.
- Travel between the two projects occurs only once a day (the engineer starts his or her day on job A, travels to job B, and then comes back home near job A).
- Overtime, if needed, is compensated at 1.5 times the regular rate (of wages only, assuming benefits are constant per hour).

The total cost ($) of the engineer per day is most nearly:

A. 968
B. 584
C. 742
D. 522

Question 1.22

Given the following, the total volume of soil (Excavation and Embankment) to be excavated (yd^3) using the Average End Method is most nearly:

Station	Area of Cross-Section (ft^2)
1+25	100
1+70	138
2+15	141
2+35	145
2+65	127

A. 635
B. 690
C. 755
D. 785

Question 1.23

A house plan is shown below. Lumber 2 x 6 studs are used on 16 in. centers in the exterior walls. The approximate number of studs required to construct the four (4) exterior walls is most nearly:

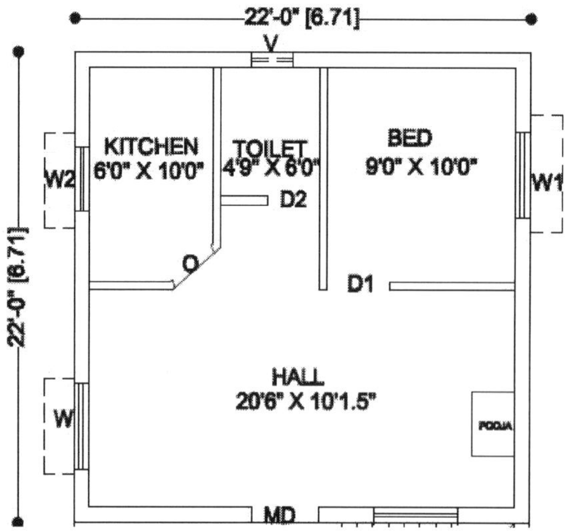

A. 66
B. 84
C. 102
D. 112

Question 1.24

A construction project has the following activities. With an imposed finish date of 16, the total float and the free-float of Activity A, respectively are most nearly:

Activity	Duration (Days)	Immediately Preceding Activity
A	3	-
B	7	A
C	4	A
D	5	B
E	8	B, C

A. Total = –2, Free = 0
B. Total = 0, Free = –2
C. Total = 0, Free = 0
D. Total = 2, Free = 2

Question 1.25

For the following land profile, the given data:

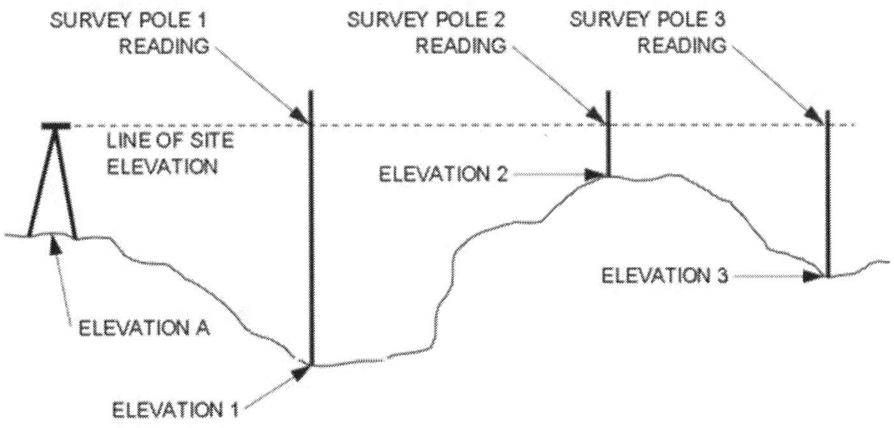

- Elevation 2 = Benchmark = 100.00 ft
- Survey Pole 1 Reading = 15.9 ft
- Survey Pole 2 Reading = 8.1 ft

The Elevation 1 (ft) is most nearly:

A. 92.2
B. 90.2
C. 124.0
D. 108.1

Question 1.26

The soil in a 1,000 ft x 10 ft x 8 ft deep trench has a swell factor of 17% and a shrinkage factor of 14%. The amount (yd^3) of backfill soil required in LCY is most nearly:

A. 564
B. 495
C. 640
D. 483

Question 1.27

A temporary scaffolding is required to support a 1,500 ft^2 roof until construction is complete.

The final roof load is 55 psf
The scaffolding load is 18 psf
The personnel/equipment load is 8 psf
The horizontal construction load is 2 psf

Use the combination: $1.2D + 1.2C_D + 1.2C_{FML} + 1.4C_{VML} + 1.6C_P + 1.6C_H + 0.5L$

The load (kip) that the scaffolding must be capable of supporting is most nearly:

A. 104
B. 124
C. 156
D. 171

Question 1.28

A mason contracted by a general contractor to build an exterior wall requiring 738 CMU blocks. The mason has to finish the wall in three days and gets paid $4.0 per block. At the end of day one, the mason has installed 220 blocks. His actual cost (including his overhead and profit) is $836.

The project is most nearly:

A. Ahead of schedule and under-budget
B. Ahead of schedule and over-budget
C. Behind the schedule and under-budget
D. Behind the schedule and over-budget

Question 1.29

If workers are doing 60 hours per week with overtime paid at 1.5 the regular rate, the productivity throughout the week is believed to decline to 80% of the regular productivity. The cost factor for labor considering this overtime is most nearly:

A. 1.5
B. 1.875
C. 1.167
D. 1.458

Question 1.30

There are 160 employees in a construction site. How many toilet seats are to be provided at the minimum level at the job site to insure occupational health?

A. 2
B. 3
C. 4
D. 5

Question 1.31

The figure below shows a survey grid for a borrow pit excavation. The number shown at each grid intersection is the depth (ft) of cut at that point. The total excavation volume (BCY or yd^3) is most nearly:

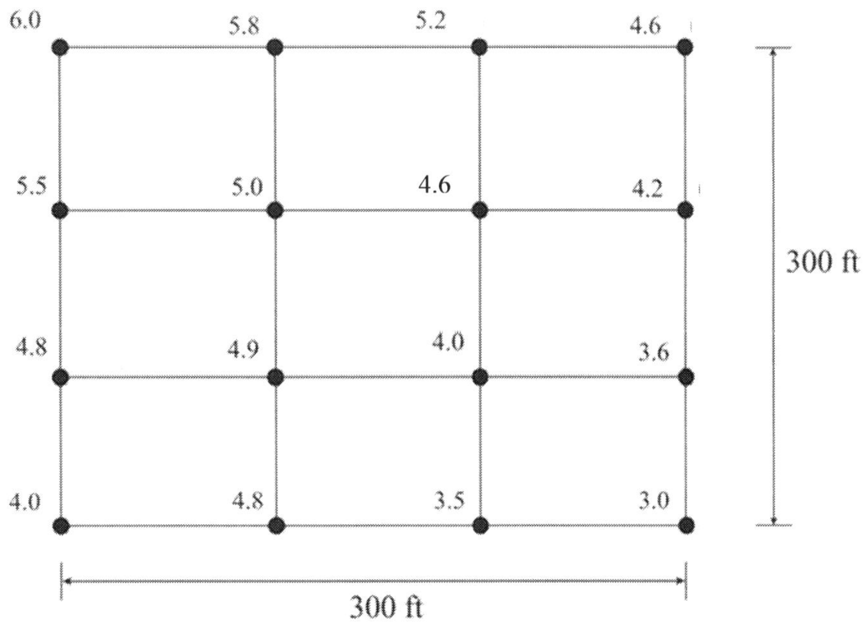

A. 15,400
B. 14,400
C. 16,400
D. 10,400

Question 1.32

When there is a 3-day lag between the end of activity A and the start of activity B, then:

A. Activity B has to wait exactly 3 days after the completion of activity A before it can start
B. Activity B has to wait at least 3 days after the completion of activity A before it can start
C. Activity B has to wait at most 3 days after the completion of activity A before it can start
D. Activity B cannot take more than 3 days in duration

Question 1.33

For a project, the structural design engineer specifies the concrete strength of 4,000 psi. Consider, the standard deviation of the concrete strength in the lab is unknown. The required average compressive strength (psi) of concrete to be prepared is most nearly:

A. 5,000
B. 4,390
C. 4,364
D. 5,200

Question 1.34

A mining area prevents access of handling equipment. A crew of two laborers at $50 per hour each and three laborers at $70 per hour each will be used to carry 200 metal pipes from the ground surface to 200-ft down the excavation. The daily work limit is 6 hours. If $8,000 is allowed over a 5-day period for the work, the minimum production rate (pipe/hr) required from the crew is most nearly:

A. 3.75
B. 5.75
C. 7.75
D. 9.75

Question 1.35

What is the difference between timber and lumber?

A. Timber is the processed wood to be used in engineering; lumber is the raw wood
B. Lumber is the processed wood to be used in engineering; timber is the raw wood
C. Timber is standing or felled trees, before they are milled into boards
D. Timber is a lumber section with dimension 5 x 5 in. or larger

Question 1.36

A concrete mix has cement : sand : aggregate = 1 : 1.5 : 2.5 by weight. The specific gravity of cement, sand, and course aggregate are 3.15, 2.62, and 2.65 respectively. The cement purchased had 94 lb in each sack. Six (6) gallons of water is used for each sack. The mix design has 4% air void by volume. The amount of cement (pcf) used in the mix is most nearly:

A. 21.3
B. 25.3
C. 29.3
D. 33.3

Question 1.37

The estimate for a four-story office building includes 48 windows per floor. It has been estimated that each window takes 2.5 carpenter man-hour to install. With an installation crew of 3 carpenters, and assuming, an 8-hour work day, the total number of days to install all the windows is most nearly:

A. 20
B. 18
C. 22
D. 25

Question 1.38

The tower crane whose capacity chart is shown in the table below is equipped with a 260-ft boom. The crane is preparing to lift a load weighing 10,000 lb. The weight of slings and the spreader bar to be used is 1,200 lb.

Table. Maximum capacity vs. lift radius for a tower crane [pounds (kilograms)]

Lift Radius ft (m)	Boom Length (maximum hook radius)—ft (m)					
	260 (79.2)	*230 (70.1)*	*200 (61.0)*	*170 (51.8)*	*140 (42.7)*	*110 (33.5)*
110 (33.5)	21564 (9781)	23607 (10708)	28458 (12908)	34857 (15811)	39680 (18000)	39680 (18000)
120 (36.6)	19584 (8883)	21465 (9737)	25938 (11765)	31842 (14444)	38097 (17281)	
130 (39.6)	17802 (8075)	19548 (8867)	23652 (10729)	29124 (13211)	34920 (15840)	
140 (42.7)	16380 (7430)	18018 (8173)	21861 (9916)	26982 (12239)		
150 (45.7)	15057 (6830)	16596 (7528)	20196 (9161)	24984 (11333)		
160 (48.8)	13699 (6214)	15143 (6869)	18534 (8407)	24705 (11206)		
170 (51.8)	12654 (5740)	14012 (6356)	17214 (7808)	23037 (10450)		
180 (54.9)	11818 (5361)	13119 (5951)	16160 (7330)			
190 (57.9)	10468 (4748)	11666 (5292)	14450 (6555)			
200 (61.0)	9700 (4400)	10811 (4904)	12440 (5643)			
210 (64.0)	9092 (4124)	10156 (4607)				
220 (67.1)	8208 (3723)	9215 (4180)				
260 (79.2)	7334 (3327)					

Minimum lift radius = 12.0 ft (3.6 m)

The maximum safe lift radius (ft) of the crane for this load is most nearly:

A. 180
B. 185
C. 190
D. 195

Question 1.39

A pump test was conducted on a test well drilled into a confined aquifer, with the results shown below. If water was pumped at a steady flow of 205 gal/min, the coefficient of permeability (ft/sec) of the permeable soil in the aquifer is most nearly:

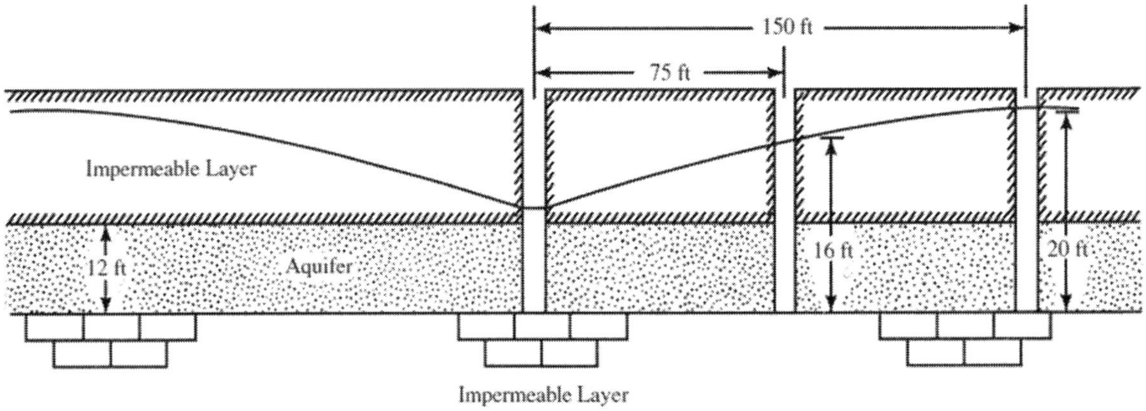

A. 0.00505
B. 0.00750
C. 0.03550
D. 0.00105

Question 1.40

Per OSHA 1926, employers with ________ or more employees must maintain records of and report occupational injuries and occupational illnesses.

[Fill up the blank with a whole number, such as 7, 97, etc.]

After taking this test, see your performance – do not get upset – work on your weak areas. Take the second exam after a few weeks – again, do not get upset – work on your weak areas again. If you score roughly 70%, the author believes that you will pass the actual exam. If you fail to pass the exam, you will lose your exam fees only – nothing else!

Solution of Practice Exam 1

Question	*Answer*	*Question*	*Answer*
1.1	B	1.21	B
1.2	B	1.22	B
1.3	D	1.23	B
1.4	B	1.24	A
1.5	C	1.25	A
1.6	A	1.26	A
1.7	D	1.27	C
1.8	C	1.28	C
1.9	B	1.29	D
1.10	B	1.30	C
1.11	D	1.31	A
1.12	A	1.32	B
1.13	C	1.33	D
1.14	C	1.34	C
1.15	C	1.35	D
1.16	B	1.36	B
1.17	C	1.37	A
1.18	B	1.38	B
1.19	D	1.39	D
1.20	A	1.40	11

Solution 1.1

AISC Manual, Chap 8:
The symbol indicates a field-placed bevel weld. Weld placed above the line indicates weld on the other side (side opposite the arrow). A bevel (groove) is made and then weld is applied.
The correct answer is B.

Solution 1.2

Taking care of unsafe condition is the contractor's responsibility.
The correct answer is B.

Solution 1.3

PE Civil Handbook: Chap 2 Construction (Section: Earthwork Construction and Layout)

For Station 0 to 1:

$$V_{cut} = L\left(\frac{A_1 + A_2}{2}\right) = (100\,\text{ft})\left(\frac{40\,\text{ft}^2 + 140\,\text{ft}^2}{2}\right) = 9{,}000\,\text{ft}^3 = \frac{9{,}000}{27}\text{yd}^3 = 333.3\,\text{yd}^3$$

$$V_{fill} = L\left(\frac{A_1 + A_2}{2}\right) = (100\,\text{ft})\left(\frac{0 + 20\,\text{ft}^2}{2}\right) = 1{,}000\,\text{ft}^3 = \frac{1{,}000}{27}\text{yd}^3 = 37\,\text{yd}^3$$

V_{cut} does not need to be adjusted for shrinkage. No swell is also mentioned.

$$V_C = \left(1 - \frac{S_h}{100}\right)V_B$$

V_{fill} adjusted for shrinkage (Bank) = 37.0 yd^3 / (1 – 0.25) = 49.3 yd^3
Compacted volume have been converted into bank volume to compare "apple to apple"
Net cut = 333.3 yd^3 – 49.3 yd^3 = 284 yd^3

The correct answer is D.

Solution 1.4

PE Civil Handbook: Chap 2 Construction (Section: Scheduling)

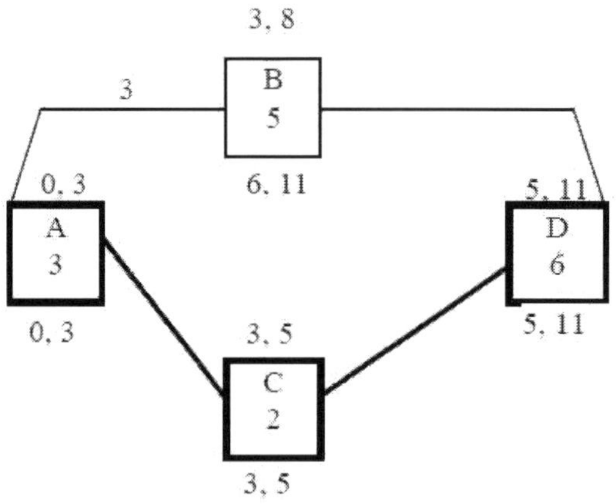

The correct answer is B.

Solution 1.5

PE Civil Handbook: Chap 2 Construction (Section: Health and Safety)

Total injuries, N = 5 + 4 + 6 = 15 (this does not include the first aid cases)

IR = N x 200,000 /T = 15 x 200,000 / 800,000 = 3.75

The correct answer is C.

Solution 1.6

No reference is required – basic estimation of construction projects.

On-center perimeter = 45 + 50 + 25 + 15 + 30 + 12 + 25 + 25 + 25 + 12 = 264 ft
Outside perimeter = 264 ft + 4 sides (0.5 ft + 0.5 ft) = 268 ft
Inside perimeter = 264 ft – 4 sides (0.5 ft + 0.5 ft) = 260 ft

Outside surface area = 268 ft x 10 ft = 2,680 ft^2
Inside surface area = 260 ft x 10 ft = 2,600 ft^2

Total surface area of wall formwork = 2,680 ft^2 + 2,600 ft^2 = 5,280 ft^2

The correct answer is A.

Solution 1.7

No reference is required – basic estimation of construction projects.

Bar 1:
Straight length = 4,000 mm + (2 x 230 mm) – (2 x 40 mm) = 4,380 mm
Cutting Length = 4,380 mm

Bar 2:
Straight length = 4,000 mm + (2 x 230 mm) – (2 x 40 mm) = 4,380 mm
Bend = 200 mm each
Cutting Length = (2 x 200 mm) + 4,380 mm = 4,780 mm

The correct answer is D.

Solution 1.8

Although it is the responsibility of design engineer to specify the configuration of reinforcement, the civil (construction) engineer should have the common knowledge that reinforcement is placed on the tension side of a reinforced-concrete section. As plain concrete is weak in tension, reinforcement is provided to increase the tensile capacity of the section.

The most efficient footing and steel will have the following:

- A larger heel than toe – to utilize the soil weight.
- Heel reinforcing near the top face – tension at top.
- Toe reinforcing near the bottom face– tension at bottom.
- Wall reinforcing placed at the face of high soil side

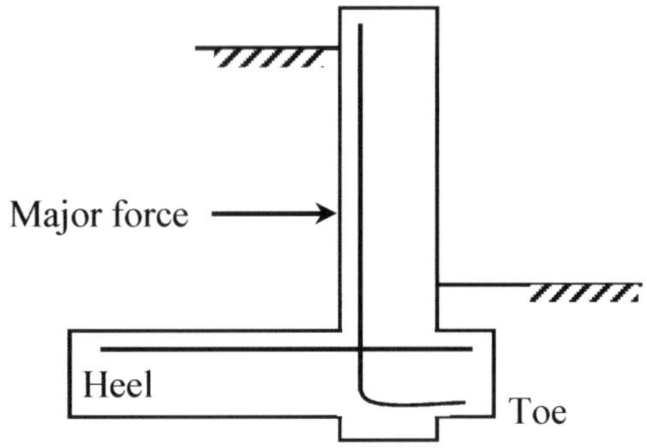

The correct (most suitable) answer is C.

Solution 1.9

No reference is required – basic estimation of construction projects.

Surface area of bottom or top 8 ft = (40 + 40 + 90 + 90) ft x 8 ft tall x 2 sides of each wall = 4,160 ft^2

Crew-hour time for bottom 8 ft = (4,160 ft^2 / 40 ft^2) = 104

Crew-hour time for top 8 ft = (4,160 ft^2 / (0.6 x 40 ft^2) = 174

Total crew-hour time = 104 + 174 = 278

The correct answer is B.

Solution 1.10

This topic is not discussed in any of the references. However, as a civil (construction) engineer, it is expected that you know the basic notations used in materials.

Two (2) means the lumber grade is No. 2. All other options are correct.

The correct answer is B.

Solution 1.11

ACI 347R *Guide to Formwork for Concrete,* 2014, Chap 4.

Table 4.2.2.1a(a)—Applicable lateral pressure equations for concrete other than SCC

Slump[*]	Internal vibration depth	Element	Rate of placement	Pressure equation
Greater than 7 in.	Any	Any	Any	4.2.2.1a(a)
Less than or equal to 7 in.	Greater than 4 ft	Any	Any	4.2.2.1a(a)
Less than or equal to 7 in.	Less than or equal to 4 ft	Column[†]	Any	4.2.2.1a(b)
		Wall[‡] less than or equal to 14 ft tall	Less than 7 ft/h	4.2.2.1a(b)
		Wall[‡] greater than 14 ft tall	Less than 7 ft/h	4.2.2.1a(c)
		Wall[‡]	7 to 15 ft/h	4.2.2.1a(c)
			Greater than15 ft/h	4.2.2.1a(a)

Eq. 4.2.2.1a(b): $C_{CP\max} = C_c C_w \left[150 + \dfrac{9{,}000R}{T}\right]$ with a minimum of $600C_w$ psf

Table 4.2.2.1a(b): Chemistry coefficient $C_c = 1.0$ for all purpose cement (Type I) and no retarder.
Table 4.2.2.1a(c): Unit weight coefficient $C_w = 1.0$ for normal-weight concrete.

Rate of placement, $R = 3.0$ ft/h
Temperature of concrete at the time of placement = 70 °F

$$C_{CP\max} = C_c C_w \left[150 + \frac{9{,}000R}{T}\right] = (1.0)(1.0)\left[150 + \frac{9{,}000(3.0)}{70}\right] = 536\,\text{psf}$$

Minimum = $600C_w = 600\,(1.0) = 600$ psf. No need to check the upper limit, Eq. 4.2.2.1a(c).

The correct answer is D.

Solution 1.12

It is just the definition of several terms used in construction engineering. No reference lists the answer directly.

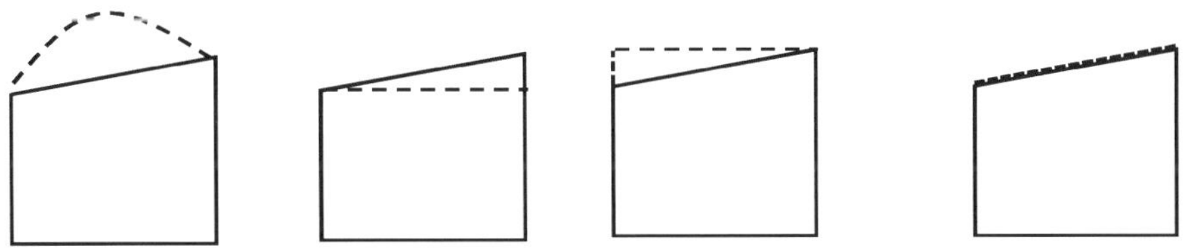

Heaped Volume: As much as possible as marked.

Water-line Volume: The maximum water level possible as marked.

Struck Volume: To the top level as marked.

Plate-line Volume: Cut surface by a plate as marked.

The correct answer is A.

Solution 1.13

PE Civil Handbook: Chap 3 Geotechnical (Section: Trench and Construction Safety) (OSHA regulation Part 1926, Subpart P, Appendix B)

For Class A soil, all simple slope excavation 20 feet or less in depth shall have a maximum allowable slope of ¾:1. This means the horizontal distance of the slope will be ¾ x 15 ft = 11.25 ft. Therefore, X = 30 ft – 11.25 ft = 18.75 ft.

For Class B soil, all simple slope excavations 20 feet or less in depth shall have a maximum allowable slope of 1:1. This means the horizontal distance of the slope will be 1 x15 ft = 15 ft. Therefore, X = 30 ft – 15 ft = 15 ft.

For Class C soil, all simple slope excavations 20 feet or less in depth shall have a maximum allowable slope of 1½:1. This means the horizontal distance of the slope will be 1½ x 15 ft = 22.5 ft. Therefore, X = 30 ft – 22.5 ft = 7.5 ft.

The correct answer is C.

Solution 1.14

No reference is required – basic estimation of construction projects.

5 sides of a brick will have mortar. The dimensions of a brick with mortar:

Width = 3.75 in. + 0.25 in. = 4.0 in.
Height = 2.25 in. + 0.25 in. + 0.25 in. = 2.75 in.
Length = 8.0 in. + 0.25 in. + 0.25 in. = 8.5 in.
Volume of a brick with mortar = 4.0 in. x 2.75 in. x 8.5 in. =93.5 in.^{3}

Volume of a brick without mortar = 3.75 in. x 2.25 in. x 8 in. = 67.5 in.^{3}

Volume of mortar in a brick = 93.5 in.^{3} – 67.5 in.^{3} = 26 in.^{3}
Total volume of mortar = 982 units x 26 in.^{3} x 1.25 = 31,915 in.^{3}
= 31,915 / 12^3 ft^3
= 18.5 ft^3

The correct answer is C.

Solution 1.15

PCA EB001, *Design and Control of Concrete Mixtures*, 17/e, 2021, Chap 13.

Table 13-1: The exposure category is F3, for freezing and thawing exposure with frequent accumulation of snow and ice with deicing chemicals.
Table 13-6: The target air content is 6.0%, for 1.0-in. aggregate and F3 exposure.

The correct answer is C.

Solution 1.16

MUTCD, 2009, Table 6C-3 and Table 6C-4:

Table 6C-3. Taper Length Criteria for Temporary Traffic Control Zones

Type of Taper	Taper Length
Merging Taper	at least L
Shifting Taper	at least 0.5 L
Shoulder Taper	at least 0.33 L
One-Lane, Two-Way Traffic Taper	50 feet minimum, 100 feet maximum
Downstream Taper	50 feet minimum, 100 feet maximum

Note: Use Table 6C-4 to calculate L

Table 6C-4. Formulas for Determining Taper Length

Speed (S)	Taper Length (L) in feet
40 mph or less	$L = \dfrac{WS^2}{60}$
45 mph or more	$L = WS$

From Table 6C-3: A shifting taper is at least 0.5L = 0.5 (300 ft) = 150 ft
Let the speed be 40 mph or less (all answer options are less than 45 mph)

From Table 6C-4: $L = \dfrac{WS^2}{60} = \dfrac{(12\,\text{ft})S^2}{60} = \dfrac{S^2}{5}$

$$\frac{S^2}{5} = 150\,\text{ft}$$

Therefore, $S = 27.4$ mph ≈ 25 mph (which is less than 40 mph – assumption valid)

The correct answer is B.

Solution 1.17

PE Civil Handbook: Chap 2 Construction (Section: Scheduling)
We can define a dummy activity as a fictitious activity inserted in an arrow network to maintain proper logic or to distinguish the activities' identities.

The correct answer is C.

Solution 1.18

No reference is required – basic analysis of construction analysis.

Diagonal of the base = $\sqrt{(30^2 + 8^2)} = 31$ ft
Half of the diagonal = 31 ft / 2 = 15.5 ft

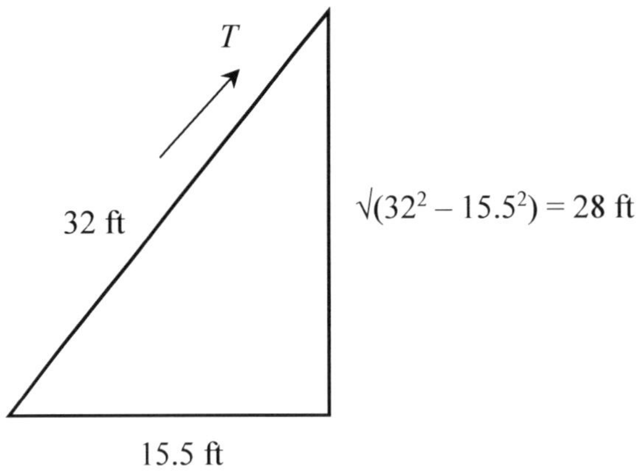

For equilibrium, sum of vertical force = 0

$$4\text{cables}\left[T\left(\frac{28}{32}\right)\right] - 30\text{ kip} = 0$$

$T = 8.57$ kip

The correct answer is B.

Solution 1.19

Both restrictions given in the diagram must be satisfied. Option D only satisfies it.
The correct answer is D.

Solution 1.20

PCA EB001, *Design and Control of Concrete Mixtures*, 17/e, 2021, pp. 52.
The correct answer is A.

Solution 1.21

No reference is required – basic estimation of construction operations.

The engineer travels between the two jobs (round trip) for no more than 2 hours (3 hours at job A + 3 hours at job B + 2 hours' travel time = 8 hours per day).
Maximum distance = 2 hours $*$ 40 mph = 80 miles round-trip or 40 miles one way. No overtime is required as it is possible to attend both projects and return within the 8 hours of a day.

Car cost = 80 miles x \$1 per mile = \$80 per day.
Total cost per day = \$63 per hour x 8 hours + \$80 = \$584

The correct answer is B.

Solution 1.22

PE Civil Handbook: Chap 2 Construction (Section: Earthwork Construction and Layout)

Volume, $V = L\left(\frac{A_1 + A_2}{2}\right)$

- V_1= (170 ft – 125 ft) x [(100 ft^2 + 138 ft^2)/2] = 5,355 ft^3
- V_2= (215 ft – 170 ft) x [(138 ft^2 + 141 ft^2)/2] = 6,278 ft^3
- V_3= (235 ft – 215 ft) x [(141 ft^2 + 145 ft^2)/2]= 2,860 ft^3
- V_4= (265 ft – 235 ft) x [(145 ft^2 + 127 ft^2)/2]= 4,080 ft^3

Total = 5,355 ft^3 + 6,278 ft^3 + 2,860 ft^3 + 4,080 ft^3 = 18,573 ft^3 $\approx$ 690 yd^3

The correct answer is B.

Solution 1.23

No reference is required – basic estimation of construction projects.

Wall length = 22 ft = 264 in.

Number of studs in a wall = 264 / 16 = 16.5 (use 16 leaving the corners empty)
Number of studs in 4 walls = 16 x 4 = 64

Number of corner = 4 (requires 2 studs each)

Number of wall openings = 6 (requires 2 extra studs each)

Total number of studs = 64 + 4 x 2 + 6 x 2 = 84

The correct answer is B.

Solution 1.24

PE Civil Handbook: Chap 2 Construction (Section: Scheduling)

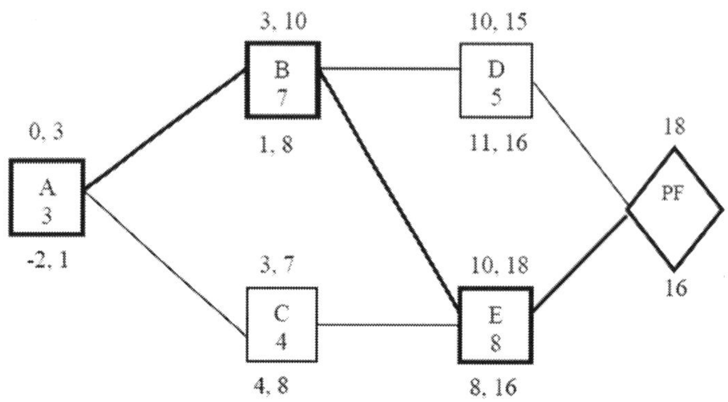

Activity	Duration	IPA	ES	EF	LS	LF	TF	FF
A	3	-	0	3	-2	1	-2	0
B	7	A	3	10	1	8	-2	0
C	4	A	3	7	4	8	1	1
D	5	B	10	15	11	16	1	1
E	8	B, C	10	18	8	16	-2	0

The correct answer is A.

Solution 1.25

PE Civil Handbook: Chap 2 Construction (Section: Earthwork Construction and Layout)

Line of Site Elevation (HI) = Benchmark (BM) + Backsight (BS)
Line of Site Elevation (HI) = 100 ft + 8.1 ft = 108.1 ft
Elevation 1 = HI – Foresight (FS)
Elevation 1 = 108.1 ft – 15.9 ft = 92.2 ft

The correct answer is A.

Solution 1.26

No reference is required – basic estimation of excavation and embankment.

Volume of trench = 1,000 ft x 10 ft x 8 ft = 80,000 ft^3
Volume after soil compaction = 80,000 ft^3 x (1 – 14%) = 68,800 ft^3
Backfill Required = Trench Volume – Compacted Volume
Backfill Required = 80,000 ft^3– 68,800 ft^3= 11,200 ft^3

Volume of compacted backfill = BCF x (1 – 14%)
BCF = 11,200 ft^3 ÷ (1 – 14%) = 13,023 ft^3

Loose volume required = Bank Volume x (1 + 17%) = 13,023 ft^3 x (1 + 17%)
= 15,237 ft^3= 15,237 ÷ 27 yd^3= 564 yd^3

The correct answer is A.

Solution 1.27

ASCE 37-14 Design Loads on Structures during Construction, Chap 2.

Given, D (dead) = 55 psf, C_D (concentrated dead) = 18 psf, C_P = 8 psf, & C_H = 2 psf.

Uniformly distributed load = $1.2D + 1.2C_D + 1.2C_{FML} + 1.4C_{VML} + 1.6C_P + 1.6C_H + 0.5L$
= 1.2(55 psf) + 1.2(18 psf) + 0 + 0 + 1.6(8 psf) + 1.6(2 psf) + 0
= 103.6 psf

Load = 103.6 psf x 1,500 ft^2 = 155,400 lb ≈ 156 kip

The correct answer is C.

Solution 1.28

PE Civil Handbook: Chap 2 Construction (Section: Scheduling)

Total number of CMUs = 738
Total Budget = 738 x \$4 = \$2,952

Without any formal analysis, this problem can be answered.
Planned production per day = 738 / 3 = 246 CMU; it produced 220 CMU; so behind the schedule.
Cost (earned) of 220 blocks = 220 x \$4 = \$880; Planned budget per day = \$2,952 / 3 = \$ 984; so, under-budget.

The correct answer is C.

The formal analysis is given below:
BCWP = Budgeted cost of work performed (Earned) = \$880
BCWS = Budgeted cost of work scheduled (Planned) = \$984
ACWP = Actual cost of work performed (Actual) = \$836
CV = *BCWP* – *ACWP* (Cost Variance = Earned – Actual) = \$880 – \$836 = \$44
SV = *BCWP* – *BCWS* (Schedule Variance = Earned – Planned) = \$880 – \$984 = – \$104
The project is behind the schedule and under-budget.

Solution 1.29

No reference is required – basic estimation of construction operations.

Pay per week = 40 hr x 1.0 + 20 hr x 1.5 = 70 regular rate
Productivity = 60 hr x 0.8 = 48 hr
Cost factor = 70 / 48 = 1.458

The correct answer is D.

Solution 1.30

Part 1926, Subpart D, Standard Number 1926.51.

One (1) toilet seat and one (1) urinal per 40 workers for number of employees of 20 or more.

Therefore, for 160 employees, 4 toilet seats are to be provided.

The correct answer is C.

Solution 1.31

PE Civil Handbook: Chap 2 Construction (Section: Earthwork Construction and Layout)

Sum of corner points = 6.0 + 4.6 + 3.0 + 4.0 = 17.6 ft
Sum of border points = 5.8 + 5.2 + 4.2 + 3.6 + 3.5 + 4.8 + 4.8 + 5.5 = 37.4 ft
Sum of interior points = 5.0 + 4.6 + 4.0 + 4.9 = 18.5 ft

The corner points are assigned a weight of 1, border points are assigned a weight of 2, and the interior points are assigned a weight of 4.

$$\text{Average depth} = \frac{(17.6\,\text{ft} \times 1) + (37.4\,\text{ft} \times 2) + (18.5\,\text{ft} \times 4)}{(4 \times 1) + (8 \times 2) + (4 \times 4)} = 4.62\,\text{ft}$$

Or, you can determine the average depth of each rectangular section, and then, the average depth of the combined section.

Area of the grid = 300 ft x 300 ft = 90,000 ft^2
Volume = 90,000 ft^2 x 4.62 ft = 415,800 ft^3 = 415,800 ft^3 / 27 = 15,400 BCY

The correct answer is A.

Solution 1.32

No reference is required – basics of scheduling.

The correct answer is B.

Solution 1.33

PCA EB001, *Design and Control of Concrete Mixtures*, 17/e, 2021, Table 13-3B.

If fewer than 15 samples are tested, or no standard deviation data is available, then for a specified concrete strength of 4,000 psi, the required strength:

$$f'_{cr} = f'_c + 1{,}200 = 4{,}000\text{ psi} + 1{,}200\text{ psi} = 5{,}200\text{ psi}$$

The correct answer is D.

Solution 1.34

No reference is required – basic estimation of construction operations.

Hourly labor cost = 2 x \$50 + 3 x \$70 = \$310
Total time = \$8,000 / \$310 per hour = 25.8 hour (less than 6 hour/day for 5 days)
Productivity rate = 200 pipes / 25.8 hour = 7.75 pipe/hour

The correct answer is C.

Solution 1.35

No reference is required – basics of woods.

Sawn lumber has different size categories, dimension lumber has a thickness up to 4 in., and timber has dimensions of 5 x 5 in. or larger. Timber has two subsections: post and timber (P&T), and beams and stringers (P&S).

The correct answer is D.

Solution 1.36

PCA EB001, *Design and Control of Concrete Mixtures*, 17/e, 2021, Chap 13.

Eq. 13-6: $V = \dfrac{\text{Mass}\,(\text{M})}{\text{Unit weight}\,(D)} = \dfrac{M}{(SG)62.4\,\text{pcf}}$

Volume of 1 sack of cement, $\dfrac{M}{(SG)62.4\,\text{pcf}} = \dfrac{94\text{ lb}}{3.15(62.4\,\text{pcf})} = 0.478\text{ ft}^3$

Volume of sand used in 1 sack of cement, $\dfrac{M}{(SG)62.4\,\text{pcf}} = \dfrac{1.5 \text{x} 94\text{ lb}}{2.62(62.4\,\text{pcf})} = 0.86\text{ ft}^3$

Volume of aggregate used in 1 sack of cement, $\dfrac{M}{(SG)62.4\,\text{pcf}} = \dfrac{2.5 \text{x} 94\text{ lb}}{2.65(62.4\,\text{pcf})} = 1.42\text{ ft}^3$

Added water in 1 sack of cement = 6 gal = 6/(7.48) ft^3 = 0.8 ft^3

Sum of cement, sand, aggregate, and water = 0.478 ft^3 + 0.86 ft^3 + 1.42 ft^3 + 0.8 ft^3
= 3.56 ft^3 (96% volume, 4% air void)

Total volume = 3.56 ft^3/ 0.96= 3.71 ft^3
Quantity of cement used = 94 lb / 3.71 ft^3 = 25.3 pcf

The correct answer is B.

Solution 1.37

No reference is required – basic estimation of construction projects.

Total number of windows = 4 x 48 = 192
Man-hour per day = 3 man x 8 hour = 24
Man-hour per window = 2.5
Window per day = 24 / 2.5 = 9.6
Total number of day = 192 / 9.6 = 20

The correct answer is A.

Solution 1.38

No reference is required – basics of crane operation and safety.

Total load = 10,000 lb + 1,200 lb = 11,200 lb
Maximum safe lift radius = 185 ft by interpolating from table.

The correct answer is B.

Solution 1.39

PE Civil Handbook: Chap 2 Construction (Section: Construction Operations and Methods)

$$Q = \frac{2\pi Kf(H-h)}{\ln\left(\frac{R_o}{r_w}\right)}$$

$$\text{or, } K = \frac{Q\ln\left(\frac{R_o}{r_w}\right)}{2\pi f(H-h)}$$

Instead of a single well, multiples wells are used in this well test.

R_o can be used as 150 ft
r_w can be used as 75 ft
H can be used as 20 ft
h can be used as 16 ft
f can be used as 12 ft

Q = 205 gpm = 205 x 0.134 ft^3/min = (205 x 0.134 / 60) ft^3/s = 0.4568 ft^3/s

$$K = \frac{\left(0.4568\,\text{ft}^3/\text{s}\right)\ln\left(\frac{150\,\text{ft}}{75\,\text{ft}}\right)}{2\pi(12\text{ ft})(20\text{ ft} - 16\text{ ft})} = 0.00105\,\text{ft}/\text{s}$$

The correct answer is D.

Solution 1.40

Standard Number 1926.35 states that for those employers with 10 or fewer employees the plan may be communicated orally to employees and the employer need not maintain a written plan.

The correct answer is 11

After taking this test, see your performance – do not get upset – work on your weak areas. Take the second exam after a few weeks – again, do not get upset – work on your weak areas again. If you score roughly 70%, the author believes that you will pass the actual exam. If you fail to pass the exam, you will lose your exam fees only – nothing else!

About the Author

Dr. M. Rashad Islam is an Associate Professor and undergraduate program coordinator at the Colorado State University Pueblo. He is also a registered Professional Engineer (PE), and an ABET Program Evaluator for Civil Engineering, Civil Engineering Technology and Construction Engineering Technology undergraduate programs. Dr. Islam received a Ph.D. degree in Civil Engineering (with the grade of 'distinction' in the research, and a GPA of 4.0 out of 4.0 in the coursework) from the University of New Mexico, USA. He has a Masters degree jointly from the University of Minho (Portugal) and the Technical University of Catalonia (Barcelona, Spain). Dr. Islam is a member of the American Society of Civil Engineers (ASCE) and is associated with the American General Contractors (AGC). Dr. Islam has more than 100 publications including several reference books, several text books, over 60 scholarly journal articles, several technical reports and conference papers.

Dr. Islam's major textbooks are:

- Pavement Design – Materials, Analysis and Highways, McGraw Hill, ISBN 9781260458916.
- Civil Engineering Materials – Introduction and Laboratory Testing, ISBN 9780367224820.
- Engineering Statics, ISBN 9781003098157.
- Construction Safety, McGraw Hill, ISBN 9781264257829.
- Engineering Dynamics - Fundamentals and Applications, ISBN: 9781032255576
- Civil Engineering PE Breadth Exams - Three Full Exams with Solutions, ISBN 9780997918090.
- Geotechnical Depth PE Civil Engineering Exams - Two Full Exams with Solutions, ISBN 9781957186009.
- Structural Depth PE Civil Engineering Exams - Two Full Exams with Solutions, ISBN 9781957186016.
- Water Resources and Environmental Depth PE Civil Engineering Exams - Two Full Exams with Solutions, ISBN 9781957186023.
- Transportation Depth PE Civil Engineering Exams - Two Full Exams with Solutions, ISBN 9781957186030.
- Construction Depth PE Civil Engineering Exams - Two Full Exams with Solutions, ISBN 9781957186047.
- FE Civil Review with 800 Solved Problems, ISBN 9780997918038.
- FE Civil Practice Exams - 2 Full Tests Including Solutions, ISBN 9780997918045.
- FE Civil Practice Problems, ISBN 9780997918083.
- FE Civil Practice Exam - Set 1: 110 Questions and Solutions, Kindle Edition
- FE Civil Practice Exam - Set 2: 110 Questions and Solutions, Kindle Edition and so on.

Practice Exam 2

Number of Questions = 40
Time = 4.0 hours

Question 2.1

The plan view of a basement foundation is shown below. The dimensions are given on center-to-center. The wall is 1-ft thick and 10-ft high. The footing is 1.5-ft wide and 3-ft high. The volume (yd^3) of concrete required without any waste consideration for the wall and footing is most nearly:

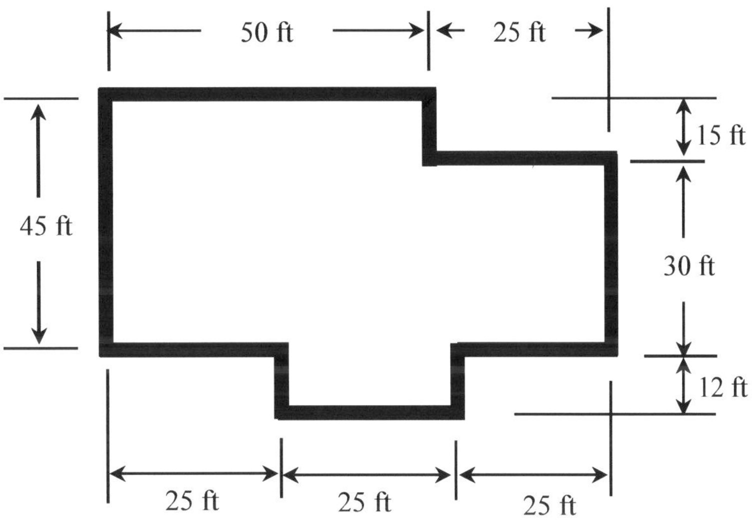

A. 132
B. 127
C. 135
D. 142

Question 2.2

The outside perimeter of a single-wythe clay masonry wall is 78 ft. The wall is 8 ft high and the units are to be laid in a running bond pattern. There are no opening in the wall. The masonry unit is standard modular unit of 8 in. long and 2-2/3 in. high including the mortar. Using a 3% percent waste factor, the total number of clay unit required to construct the wall is most nearly:

A. 4,340
B. 4,220
C. 4,120
D. 4,440

Question 2.3

Profiles of two sections of a 100-ft length of a highway is given below. The shrinkage of the soil is 10%.

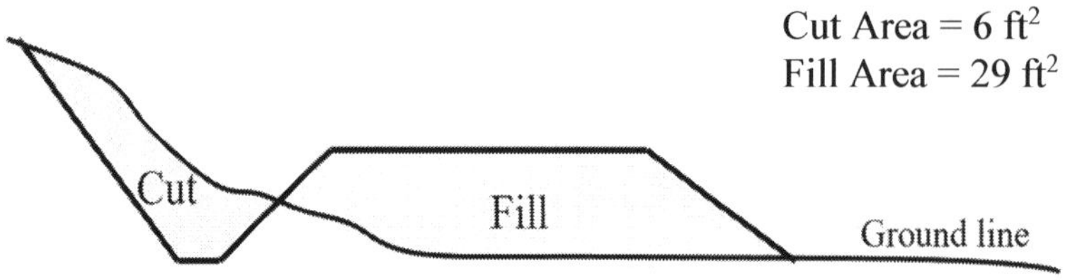

Section 1

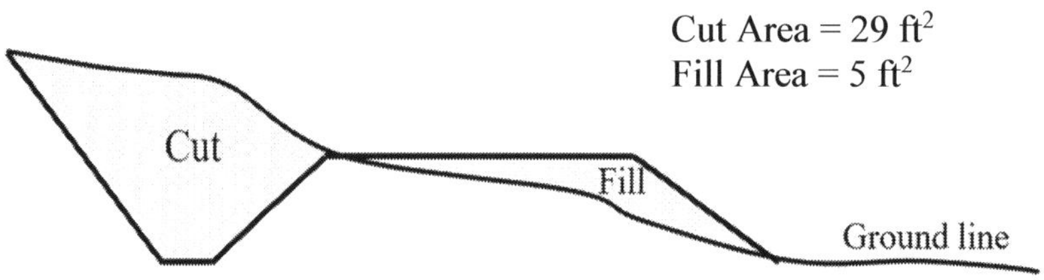

Section 2

The required cut or fill required between stations 1 and 2 is most nearly:

A. 4.5 yd^3 fill
B. 1.8 yd^3 fill
C. 4.5 yd^3 cut
D. 1.8 yd^3 cut

Question 2.4

The critical path of a project is 232 days long. The next longest path is 228 days. We cut the critical path by 6 days. The project duration now is most nearly:

A. 226 days.
B. 228 days.
C. 226 or 228 days, depending on whether the shortened activities are on the 2nd longest path.
D. 231 days.

Question 2.5

A wood frame house being constructed with a hip roof with a pitch of 6/12. The roof plan view is given below. Consider a waste factor of 12% and each piece of plywood sheet measures 8 ft by 4 ft. The number of plywood sheets required is most nearly:

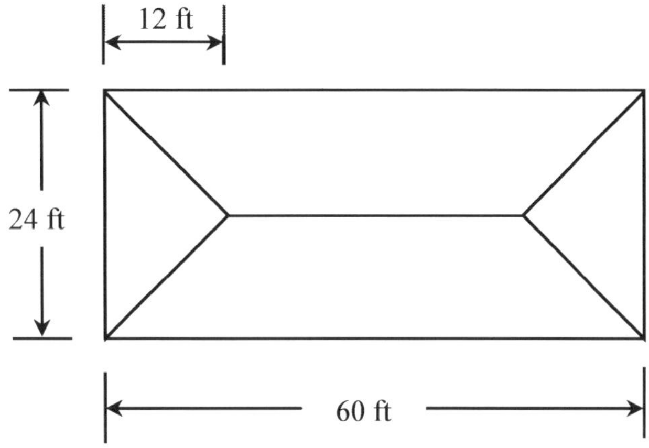

A. 52
B. 57
C. 65
D. 49

Question 2.6

A concrete building requires 50 concrete beams with 12 in. by 24 in. cross-section. The beams are 15-ft long. The materials required per 100 square foot contact area (SFCA) considering some surplus and the unit cost is listed below.

	Amount	Unit Cost
Plywood (sft)	100	$1.00
Lumber (bf)	300	$1.40
Nails (lb)	1.5	$1.2
Ties	15	$0.75

The total cost ($) of the materials for the formwork is most nearly:

A. 20,000
B. 18,000
C. 22,000
D. 17,650

Question 2.7

A bracket is anchored to a concrete wall using a bolt screwed into an imbedded insert as shown below. The developed tension (kip) in the bolt is most nearly:

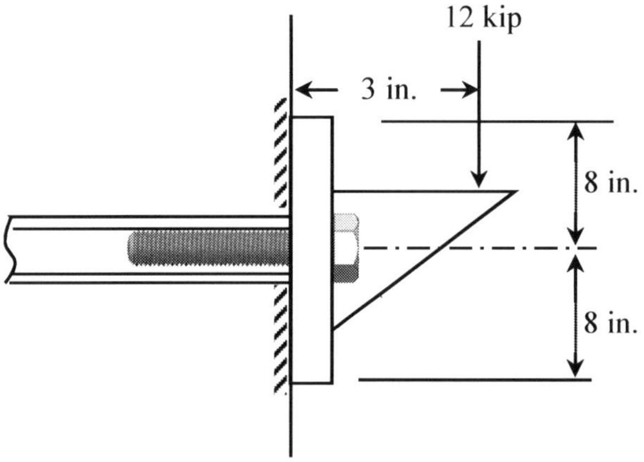

A. 0
B. 4.5
C. 12
D. 32

Question 2.8

A construction equipment company sells a tool whose selling price is $10. Its fixed cost and the variable expenses per piece are given below. The minimum number of tools it needs to sell per month to reach the break-even point is most nearly:

Fixed Costs per Month		**Variable Costs per Tool**	
General Labor	$1,500	Paint	$0.50
Rent	$3,000	Package	$0.05
Insurance	$200	Tag	$0.01
Advertising	$500	Material 1	$3.00
Utilities	$450	Material 2	$2.00
Total	$5,650	Total	$5.56

A. 828
B. 970
C. 1,000
D. 1,273

Question 2.9

A soil has a swell value of 12%. If 100 BCY (bank cubic yard) of excavation is made, what will be the volume of the spoil pile in LCY (loose cubic yard)?

A. 112 BCY
B. 112 LCY
C. 88 BCY
D. 88 LCY

Question 2.10

While excavating, a contractor hits a rock layer. Since the plans and soil report did not mention such rock, the contractor may file a claim under:

A. Force majeure
B. Differing site conditions
C. Design errors/omissions
D. Unusual weather conditions

Question 2.11

A concrete slab is being constructed for a warehouse building. When the concrete mix is placed on a hot windy day, it is not permissible to:

A. Keep the mix water-cool and aggregate moist by shading and sprinkling
B. Add field water as needed to obtain the desired consistency and workability
C. Spray the concrete surface or protect them with wet burlap to retard hardening
D. Moisten the forms and the reinforcing before placing the concrete to minimize the evaporation

Question 2.12

There are two (2) $L3 \times 3 \times \frac{1}{4}$ sections in the following welded connection. The welding symbol shows the welding specification. The total welding length (in.) is most nearly:

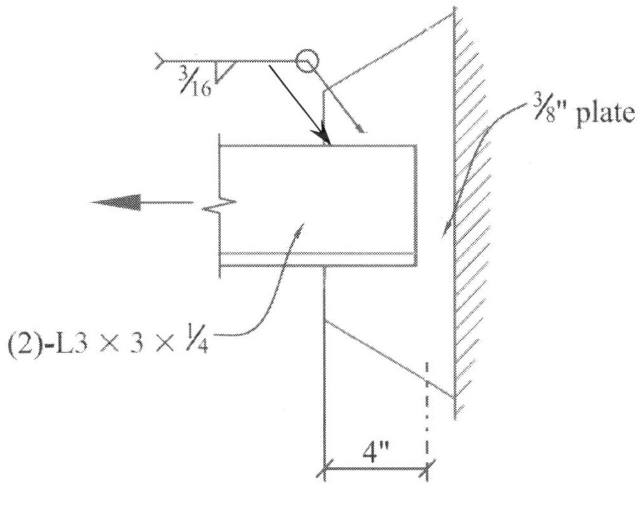

A. 22
B. 11
C. 8
D. 10

Question 2.13

The mechanical grading, "2100Fb-1.8E" on a lumber section means –

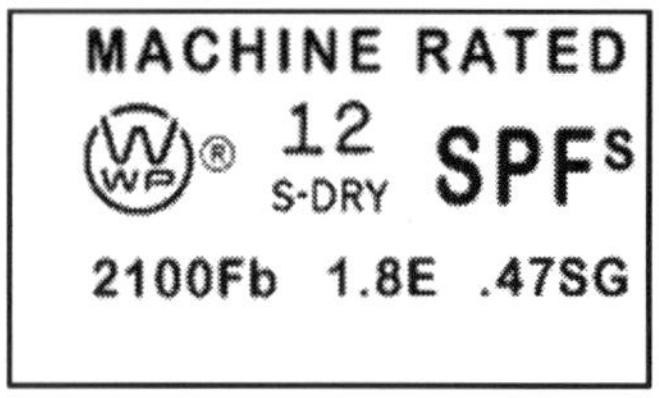

A. Its compressive strength is 2,100 psi
B. Its modulus of elasticity is 1.8 ksi
C. Its modulus of elasticity is 1,800,000 psi
D. It is just a symbolic number

Question 2.14

Materials are best managed with the concept of:

A. Buying everything in advance to make sure it is available when needed
B. Buy and deliver at installation time so we don't need to store anything
C. Minimize overall cost by taking a balanced approach between A and B
D. None of the above

Question 2.15

Wood strut is commonly used to support a roadside signboard as shown below. At this stage, the wind exerts a force of 280 lb/ft at the height of 14 ft from the ground. The self-weight of the signboard is 400 lb/ft. The foundation of the signboard can be considered a hinge support. The developed force (lb/ft) in the wood strut if the system is in equilibrium is most nearly:

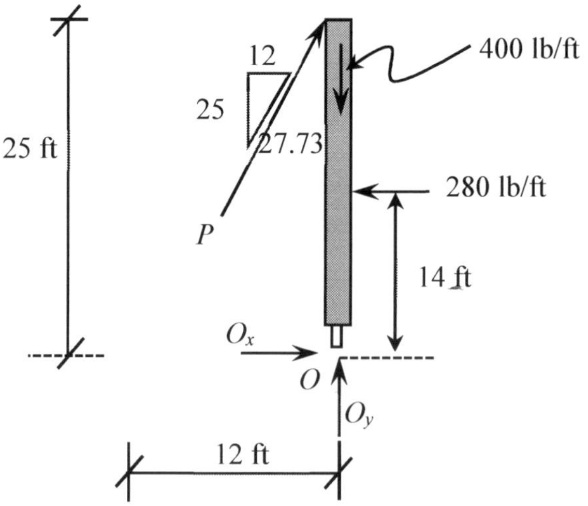

A. 322
B. 338
C. 362
D. 392

Question 2.16

The front view of the following image is most nearly:

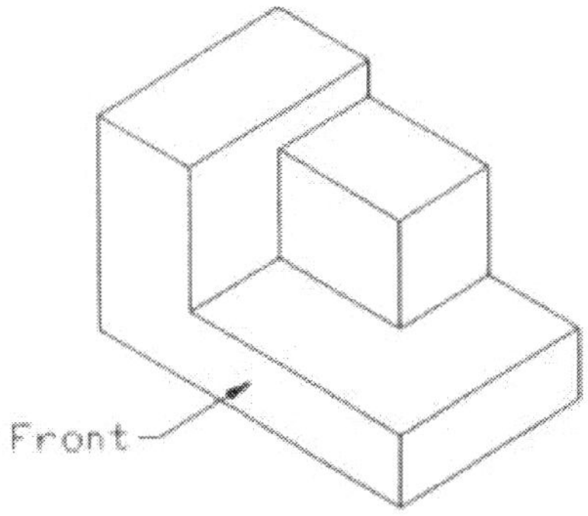

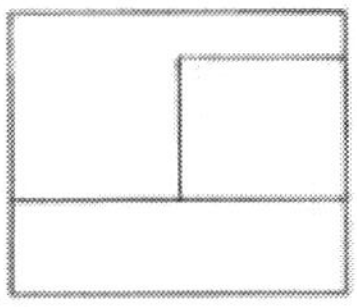

A.

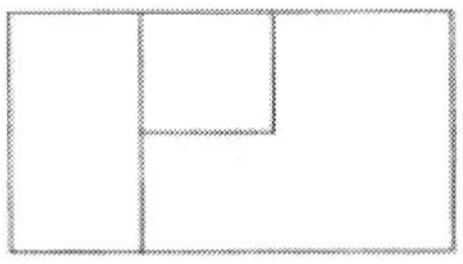

B.

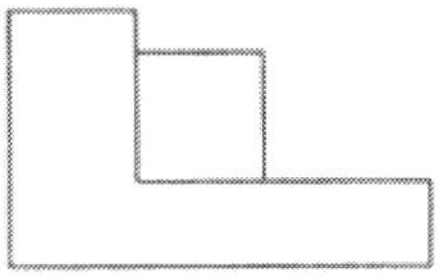

C.

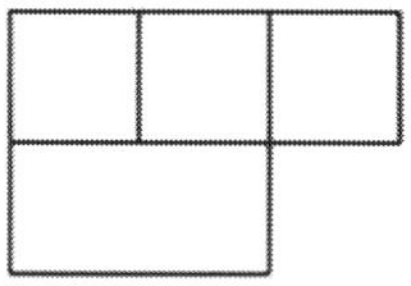

D.

Question 2.17

A construction project has the following 5 activities. With an imposed finish date of 29, the free-floats of Activities A, C and E, respectively are most nearly:

Activity	Duration (Days)	Immediately Preceding Activity
A	7	-
B	10	-
C	15	A
D	8	A, B
E	4	C, D

A. A = 0, C = 0, E = 3
B. A = 3, C = 3, E = 3
C. A = 0, C = 3, E = 3
D. A = 3, C = 0, E = 3

Question 2.18

The 13 carcinogens does not include:

A. Hydrogen Sulfide
B. Benzidine
C. 4-Aminodiphenyl
D. Ethyleneimine

Question 2.19

A construction project has the following activities with their durations and relationships shown. For the Activity B, the total float and free float, respectively are most nearly:

Activity	Duration (Days)	Immediately Preceding Activity	Type of Relationship	Lag
A	3	–		
B	5	A	Start-Start	3
C	2	A		
D	6	B	Finish-Finish	
		C		

A. Total = 0, Free = 0
B. Total = 3, Free = 0
C. Total = 0, Free = 3
D. Total = 3, Free = 3

Question 2.20

A site is to be leveled to construct a roadway. The active section is 200 feet long and has the following station areas:

Station	Area of Cross-Section (Cut)	Area of Cross-Section (Fill)
0+00	150	72
0+50	102	87
1+00	110	105
1+50	96	100
2+00	81	127

Using the average end area method, the net soil surplus/deficit required for this site is most nearly:

A. 60 yd^3 of surplus
B. 90 yd^3 of surplus
C. 60 yd^3 of deficit
D. 30 yd^3 of deficit

Question 2.21

Samples of coarse aggregate from a stockpile are brought to the laboratory, where the following data are found:

Mass of aggregate submerged in water = 1,515 g
Mass of oven-dry aggregate = 2,165 g
Mass of SSD aggregate = 2,354 g

The percent absorption of the aggregate is most nearly:

A. 8.7
B. 9.3
C. 10.3
D. 9.9

Question 2.22

A flooring subcontractor has to install three types of flooring to an existing building:

Type	Quantity	Unit Price	Total Price
Carpet	2,600 SF	$2.30 /SF	$ 5,980
Vinyl tile	1,200 SF	$2.95 /SF	$ 3,540
Ceramic tile	3,330 SF	$4.95 /SF	$16,484

The subcontractor was given 10 days to finish the work. Assume he has crews to start all three jobs concurrently. Four days after the start of the work, he found out:

Type	Quantity Installed	Actual Cost
Carpet	1500 SF	$3,150
Vinyl tile	800 SF	$2,488
Ceramic tile	750 SF	$3,638

The installation of carpet after this 4-day is most nearly:

A. under budget, and ahead of schedule
B. over budget, and ahead of schedule
C. under budget, and behind the schedule
D. over budget, and behind the schedule

Question 2.23

Soil from a borrow pit to be used for construction of an embankment gave the following laboratory results when subjected to the ASTM D 698 Standard Proctor test:

- Maximum dry unit weight = 118.5 lb/ft^3
- Optimum moisture content = 12.5%

During the construction of the soil embankment, the contractor achieved the following:

- Dry unit weight reached by field compaction = 111.8 lb/ft^3
- Actual water content = 13.7%

The percent of compaction achieved by the contractor is most nearly:

A. 94
B. 92
C. 96
D. 90

Question 2.24

The preferred cross sections for channels with abrupt slope changes is shown below. This chart is applicable to all Vee ditches, rounded channels with a bottom width less than 2.4 m [8 ft] and trapezoidal channels with bottom widths less than 1.2 m [4 ft].

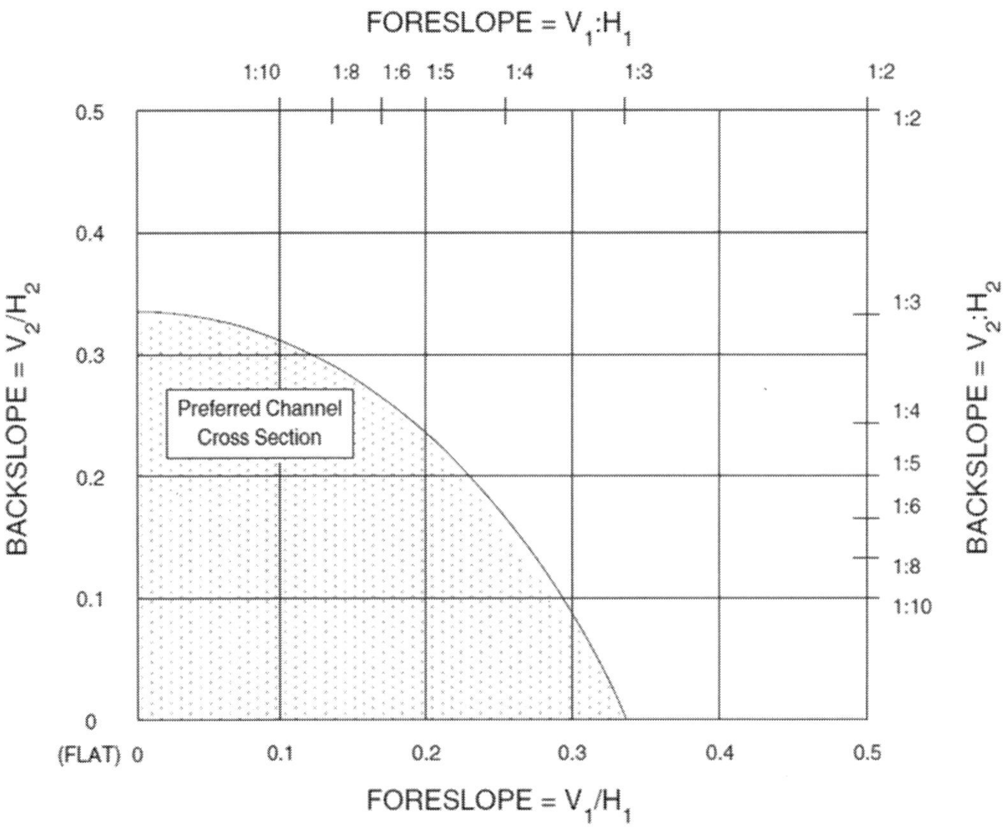

Based on the criteria provided, the steepest backslope ($V_2 : H_2$) preferred in this ditch shown below is most nearly:

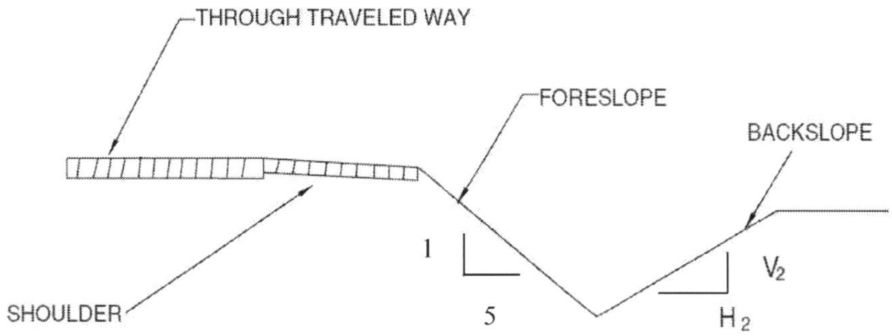

A. 1:2
B. 1:3
C. 1:5
D. 1.6

Question 2.25

A city plans to place a 5 ft x 5 ft sculpture at the middle of a 4-lane street. The speed limit of the street is 50 mph. The recommended merging taper length (ft) for the sculpture is most nearly:

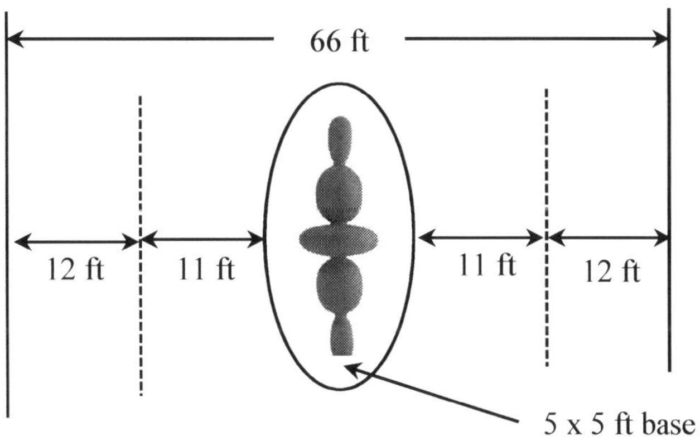

A. 500
B. 550
C. 390
D. 420

Question 2.26

Based on PCA 2016 below, the required volumes (m^3) of components required to make a batch of 0.20 m^3 of an air-entrained concrete mix with the NMAS of 12.5 mm is most nearly:

Table. Relative Components of Concrete for Small Jobs by Bulk Volume

NMAS in. (mm)	Air-Entrained Concrete			
	Cement	Wet Fine Aggregate	Wet Coarse Aggregate	Water
3/8 (9.5)	0.190	0.429	0.286	0.095
1/2 (12.5)	0.174	0.391	0.348	0.087
3/4 (19)	0.160	0.360	0.400	0.080
1.0 (25)	0.154	0.346	0.423	0.077
1.5 (37.5)	0.148	0.333	0.445	0.074
*The combined volume is approximately 2/3 of the sum of the original bulk volumes.				

Adapted from Kosmatka, Steven H. and Wilson, Michelle L., 2016. Design and Control of Concrete Mixtures, EB001, 16th edition, Portland Cement Association, Skokie, IL, Table 15.17.

A. Fine aggregate = 0.1044 and coarse aggregate = 0.1173
B. Cement = 0.0522 and coarse aggregate = 0.1173
C. Coarse aggregate = 0.1044 and water = 0.1173
D. Cement = 0.0522 and water = 0.0261

Question 2.27

Free float represents:

A. The portion of total float where it is not influenced by preceding activities
B. The portion of total float where it does not influence following activities
C. The waiting period which does not affect the total duration of the project
D. None of the above

Question 2.28

The lift study indicates that a crane imposes a maximum load of 48 tons on the outrigger. The permissible ground pressure is 200 kPa. The required size (m^2) of the crane mat is most nearly:

A. 2.00
B. 2.35
C. 4.70
D. 1.18

Question 2.29

Referring to the grade profile and mass diagram for a roadway construction project, which of the following is/are true?

[select all that apply]

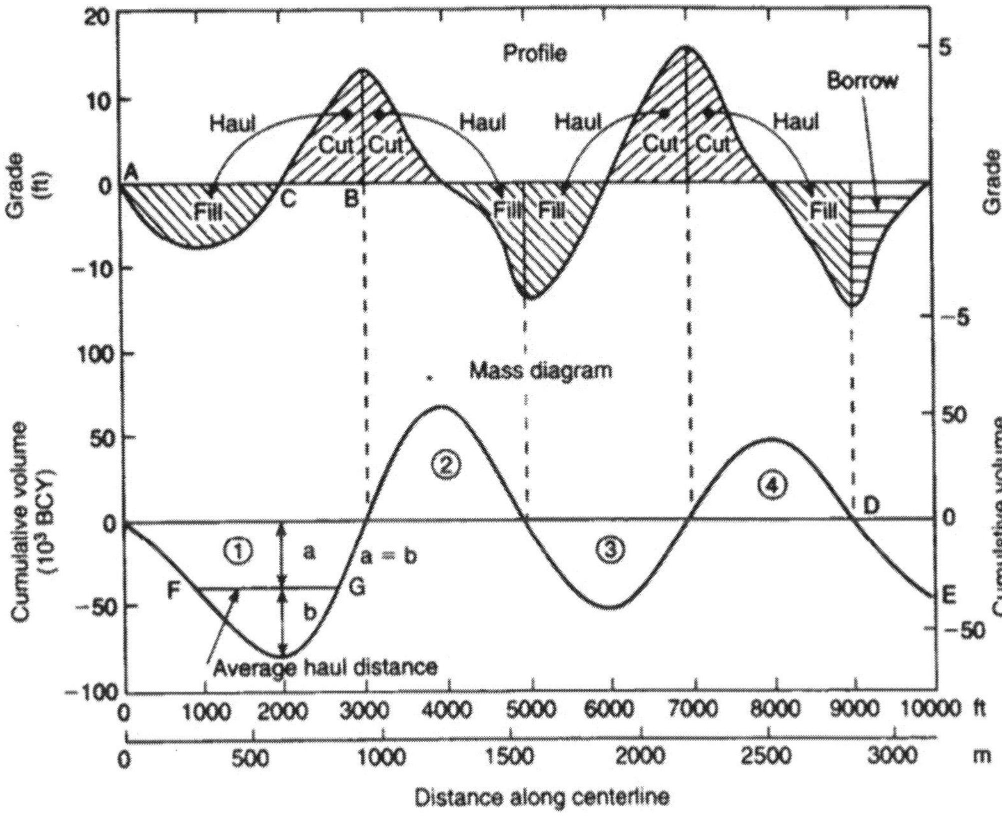

A. The point C on the profile corresponding to the lowest point of the mass diagram within section 1. This is the point at which the excavation changes from fill to cut.

B. Since the mass diagram has a negative value from point D to the end, the ordinate at point E (# 50,000 BCY or # 38,230 BCM) represents the volume of material which must be brought in from a borrow pit to complete the roadway embankment.

C. The approximate length of the line F-G represents the average haul distance for section 1, which is 1,800 ft or 549 m.

Question 2.30

The maintenance on a machine is expected to be $155 at the end of the first year, and increasing $35 each year for the following seven years. The interest rate is 6%. The present sum of money ($) that would need to be set aside now to pay the maintenance for the eight-year period is most nearly:

A. 1,560
B. 1,660
C. 1,760
D. 1,860

Question 2.31

A formal CPM analysis for a project shows the planned costs to date as $95,000 and the charge to the job as $100,000. If the reported earned value of the work to date is found to be $80,000, the project is most nearly:

A. Ahead of schedule and under-budget
B. Ahead of schedule and over-budget
C. Behind the schedule and under-budget
D. Behind the schedule and over-budget

Question 2.32

The estimate for a three-story hotel includes 24 plumbing fixtures on the first floor and 36 plumbing fixtures on each of the second and the third floors. It is estimated that each fixture takes one plumber 1.5 hours to install. With an installation crew of 2 plumbers and assuming an 8-hour work day, the total number of days required to install all the plumbing fixtures is most nearly:

A. 18
B. 9
C. 4.5
D. 5.6

Question 2.33

In a concrete mix design, a student is proposing 70% aggregate, 20% cement, and 10% water. If the amount of aggregate used is 20 kg, the amount of cement (kg) is most nearly:

A. 0.175
B. 5.71
C. 2.86
D. 8.97

Question 2.34

A power-shift transmission crawler tractor has a rated blade capacity of 10 LCY (loose cubic yard). The common dozer fixed time and operating speeds are provided below. The dozer is excavating loose common earth and pushing it a distance of 200 ft. Maximum reverse speed in third range is 5 mph. The total cycle time (time for a single disposing of earth) in minute of the dozer is most nearly:

Table. Dozer fixed time

Operating Conditions	Time (min)
Power-shift transmission	0.05
Direct-drive transmission	0.10
Hard digging	0.15

Table. Dozer operating speeds (dozing and return)

Operating Conditions	Speeds
Dozing	
Hard materials, haul 100 ft (30 m) or less	1.5 mi/h (2.4 km/h)
Hard materials, haul over 100 ft (30 m)	2.0 mi/h (3.2 km/h)
Loose materials, haul 100 ft (30 m) or less	2.0 mi/h (3.2 km/h)
Loose materials, haul over 100 ft (30 m)	2.5 mi/h (4.0 km/h)
Return	
100 ft (30 m) or less	Maximum reverse speed in second range (power shift) or reverse speed in gear used for dozing (direct drive)
Over 100 ft (30 m)	Maximum reverse speed in third range (power shift) or highest reverse speed (direct drive)

A. 0.91
B. 0.45
C. 1.41
D. 1.59

Question 2.35

A hypothetical construction project has five activities (A to E). Their durations and immediately preceding activity (IPA) are shown. The correct node diagram is most nearly:

Activity	Duration	IPA
A	3	-
B	7	A
C	4	A
D	5	B
E	8	B, C

A.

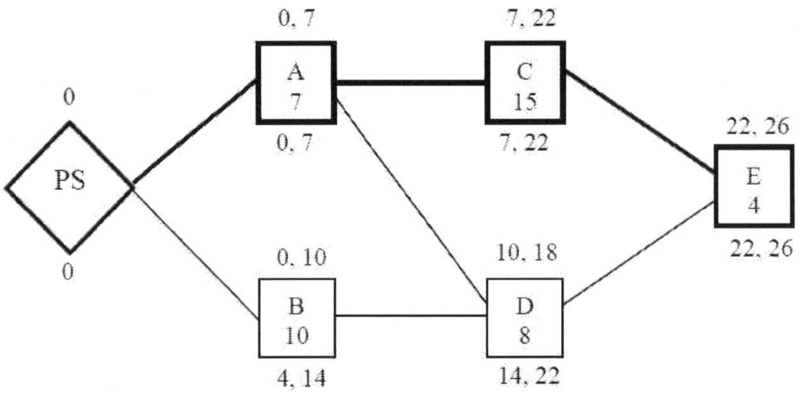
0
PS
0
0, 7
A
7
0, 7
7, 22
C
15
7, 22
22, 26
E
4
22, 26
0, 10
B
10
4, 14
10, 18
D
8
14, 22

B.

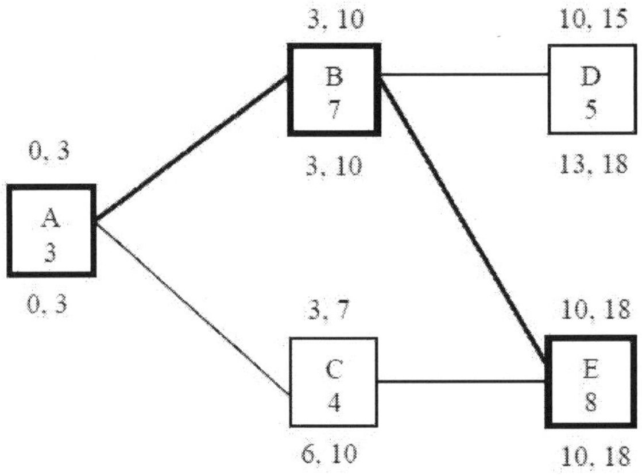
0, 3
A
3
0, 3
3, 10
B
7
3, 10
10, 15
D
5
13, 18
3, 7
C
4
6, 10
10, 18
E
8
10, 18

C.

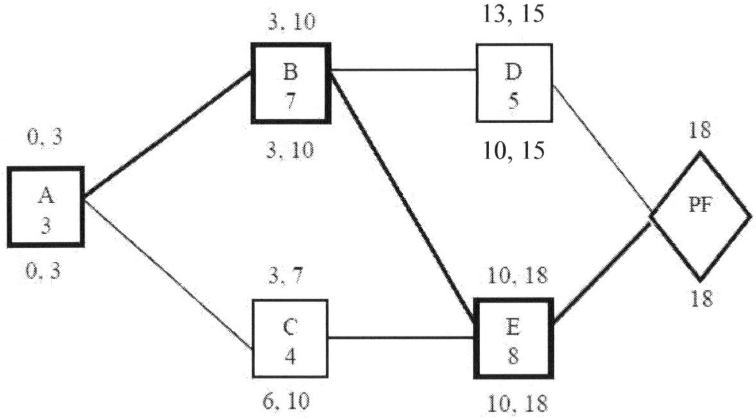
0, 3
A
3
0, 3
3, 10
B
7
3, 10
13, 15
D
5
10, 15
3, 7
C
4
6, 10
10, 18
E
8
10, 18
18
PF
18

D.

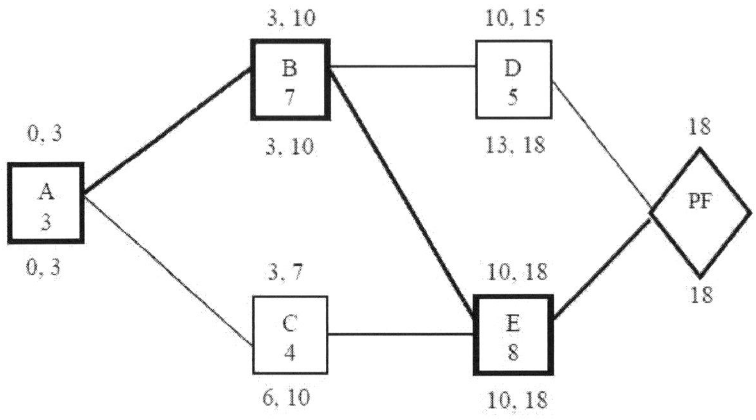
0, 3
A
3
0, 3
3, 10
B
7
3, 10
10, 15
D
5
13, 18
3, 7
C
4
6, 10
10, 18
E
8
10, 18
18
PF
18

Question 2.36

The column and girder sizes of a two-story moment frame of a temporary support structure are shown below. Assume the frame as two-dimensional, meaning there is no members perpendicular to the page. For a W-section, the first number represents the nominal depth (in.) and the second number represents the self-weight (lb/ft). For example, W 18 x 40 means a wide flange section with the nominal depth of 18 in. and the self-weight of 40 lb/ft. Point *C* will carry 50% of the girders' load from each side and the columns loads.

The total dead load (kip) at *C* considering the self-weight of the members is most nearly:

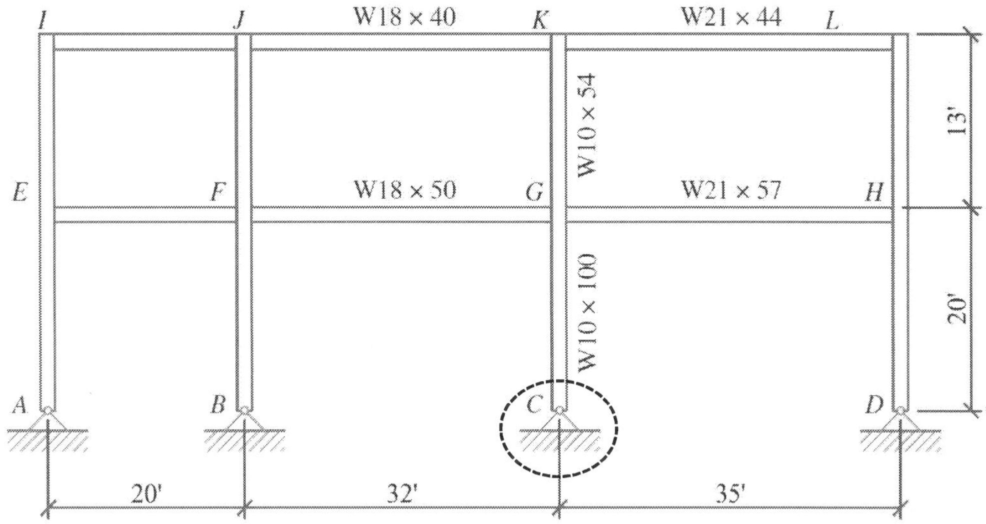

A. 5.91
B. 9.11
C. 5.11
D. 8.11

Question 2.37

A ½-in.-diameter, 7-wire prestressing strand is to be stressed in the pretensioning bed shown with a jacking force of 14 tons. The length from the dead-end anchorage to the back of the live-end hydraulic jack is 310 ft. For the strand properties indicated, the expected strand elongation (in.) is most nearly:

$E_{ps} = 27{,}500$ ksi
$A_{ps} = 0.153$ in.^{2}

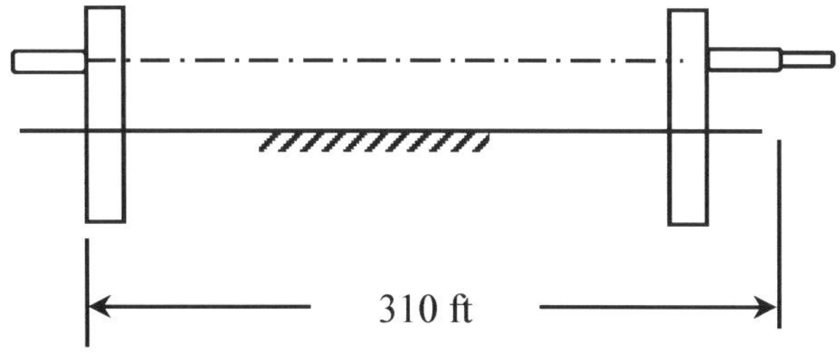

A. 3.5
B. 12.3
C. 19.3
D. 24.7

Question 2.38

The cracking of concrete pavement caused by freeze-thaw of aggregate is called:

A. Plastic shrinkage cracking
B. Early-age cracks
C. Shrinkage cracks due to improper curing
D. Durability cracking

Question 2.39

A self-propelled tamping foot roller is operating under the following conditions:

- Average speed = 5 mi/h
- Compacted lift thickness = 6 in.
- Effective roller width = 10 ft
- Job efficiency = 0.75
- Number of passes = 4

The production in compacted cubic yards per hour is most nearly:

A. 917
B. 548
C. 458
D. 588

Question 2.40

A soil weighs 1,960 lb/yd^3 in loose condition, 2,800 lb/yd^3 in natural state, and 3,500 lb/yd^3 after compaction. How many compacted cubic yards (CCY) is contained in 1.0 million loose cubic yards of this soil?

A. 1,786,000
B. 700,000
C. 875,000
D. 560,000

If you have any question regarding a problem or any other issue regarding the PE Civil Construction Depth Exam, please feel free to email the author at islamunm@gmail.com

Solution of Practice Exam 2

Question	*Answer*	*Question*	*Answer*
2.1	D	2.21	A
2.2	A	2.22	A
2.3	A	2.23	A
2.4	C	2.24	C
2.5	B	2.25	A
2.6	A	2.26	D
2.7	B	2.27	B
2.8	D	2.28	B
2.9	B	2.29	A, B, C
2.10	B	2.30	B
2.11	B	2.31	D
2.12	A	2.32	B
2.13	C	2.33	B
2.14	C	2.34	C
2.15	C	2.35	D
2.16	C	2.36	A
2.17	A	2.37	D
2.18	A	2.38	D
2.19	D	2.39	A
2.20	A	2.40	D

Solution 2.1

No reference is required – basic estimation of construction projects.

On-center perimeter = 45 + 50 + 25 + 15 + 30 + 12 + 25 + 25 + 25 + 12 = 264 ft
Cross-section of wall and footing = (1 ft x 10 ft) + (1.5 ft x 3 ft) = 10 ft^2 + 4.5 ft^2 = 14.5 ft^2
Total volume = 264 ft x 14.5 ft^2 = 3,828 ft^3 = 3,828 /27 yd^3 = 142 yd^3

The correct answer is D.

Solution 2.2

No reference is required – basic estimation of construction projects.

Surface area of wall = 78 ft x 8 ft = 624 ft^2
Surface area of one standard unit = 8 in. x 2-2/3 in. = 21.36 $in.^2$ = 0.148 ft^2
Number of units required = 624 ft^2 / 0.148 ft^2 = 4,216
Number units required considering 3% waste = 1.03 x 4,216 ≈ 4,340

The correct answer is A.

Solution 2.3

PE Civil Handbook: Chap 2 Construction (Section: Earthwork Construction and Layout)

$$V_{cut} = L\left(\frac{A_1 + A_2}{2}\right) = (100\,\text{ft})\left(\frac{6\,\text{ft}^2 + 29\,\text{ft}^2}{2}\right) = 1{,}750\,\text{ft}^3 = \frac{1{,}750}{27} = 64.8\ \text{yd}^3$$

$$V_{fill} = L\left(\frac{A_1 + A_2}{2}\right) = (100\,\text{ft})\left(\frac{29\,\text{ft}^2 + 5\,\text{ft}^2}{2}\right) = 1{,}700\,\text{ft}^3 = \frac{1{,}700}{27} = 63.0\ \text{yd}^3$$

Shrinkage for fill = 63.0 yd^3 x 0.1 = 6.3 yd^3
Total fill = 63.0 yd^3 + 6.3 yd^3 = 69.3 yd^3

Total cut and fill between stations 1 and 2 = 69.3 yd^3 fill – 64.8 yd^3 cut = 4.5 yd^3 fill
Note: no allowance is made for expansion as per the given data.

The correct answer is A.

Solution 2.4

No reference is required – basics of scheduling.

The correct answer is C.

Solution 2.5

No reference is required – basic estimation of construction projects.

The roof pitch is 6/12, meaning, the roof height is 6 ft for a 12-ft distance to the center of the roof.

The slope length of one side of the roof = $\sqrt{(12\,\text{ft})^2 + (6\,\text{ft})^2} = 13.4\,\text{ft}$

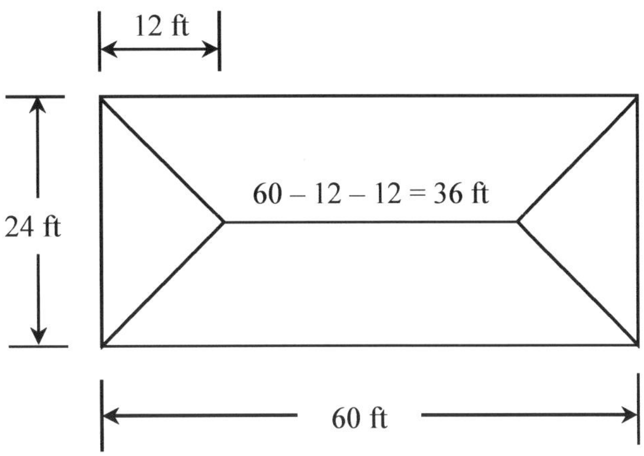

Area of the roof = 2 [½ (60 ft + 36 ft) x 13.4 ft + ½ x 24 ft x 13.4 ft] = 1,608 ft^2
Required sheet after 12% waste = 1.12 x 1,608 ft^2 = 1,800 ft^2
Area of a sheet = 8 ft x 4 ft = 32 ft^2
Number of sheet = 1,800 ft^2 / 32 ft^2 = 56.25 ≈ 57

The correct answer is B.

Solution 2.6

No reference is required – basic estimation of construction projects.

Forms are used on sides and bottom. The top is kept open for pouring concrete.

Total contact area = 50 beams x (2 sides and 1 bottom) x 15 ft long
= 50 beams x (2 ft + 2ft + 1 ft) x 15 ft long
= 3,750 SFCA

	Amount	Unit Cost	Item cost
Plywood (sft)	100	$1.00	(3,750/100) x 100 sft x $1.00 = $3,750
Lumber (bf)	300	$1.40	(3,750/100) x 300 bf x $1.40 = 15,750
Nails (lb)	1.5	$1.20	(3,750/100) x 1.5 lb x $1.20 = $68
Ties	15	$0.75	(3,750/100) x 15 x $0.75 = $422
Total			= $19,990 ≈ $ 20,000

The correct answer is A.

Solution 2.7

No reference is required – basic analysis of construction equipment and methods.

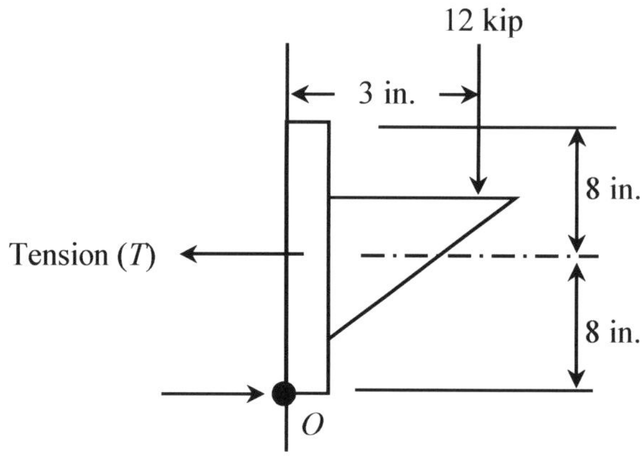

$\sum M$ at $O = 0$
12 kip (3 in.) – T (8 in.) = 0
$T = 4.5$ kip

The correct answer is B.

Solution 2.8

PE Civil Handbook: Chap 1 General Engineering (Section: Engineering Economics)

Variable Costs per tool = \$5.56
Selling price per tool = \$10.00
Revenue per tool = \$10 – \$5.56 = \$4.44
Total fixed cost per month = \$5,650
Number of tools to cover this = \$5,650/\$4.44 = 1,272.5 ≈ 1,273

The correct answer is D.

Solution 2.9

PE Civil Handbook: Chap 2 Construction (Section: Earthwork Construction and Layout)

The spoil pile is loose and larger in volume compared to the bank (in situ soil).

LCY = 100 BCY x 1.12 = 112 LCY.

The correct answer is B.

Solution 2.10

No reference is required – basic construction management issues.

Rock layer is a site condition which may not be known before the construction stats. Force majeure is unexpected natural calamities such as storm, earthquake, or flood etc.

The correct answer is B.

Solution 2.11

No reference is required although PCA EB001 (2011) discusses the topics.

Beyond the mix design, no further water can be added in the mix. The correct answer is B.

Solution 2.12

AISCM, Table 8.2 shows the round (O) welding symbol means that the welding will be all around the L-section (2 sides of 4 in. each and 1 end of 3 in.). There are 2 sections. Welding length = 2 sides (4 in. + 4 in. + 3 in.) = 22 in.

The correct answer is A.

Solution 2.13

No reference is required – basics of woods.

The mechanical grading, "2100Fb-1.8E" means its reference flexural/bending strength value is 2,100 psi and its modulus of elasticity is 1.8 Mpsi, or 1,800,000 psi.

The correct answer is C.

Solution 2.14

No reference is required – basic resource management tasks.

For the first option, storage cost may be huge for large and vulnerable materials. For the second option, there is a risk that material may not arrive on time leaving labor no work. The combination should yield better results.

The correct answer is C.

Solution 2.15

No reference is required – basic analysis of construction operations.

To be equilibrium, the summation of moment about the point O must be zero.

$$\sum M \text{ at } O = 0 \; (\circlearrowright + ve)$$

$$P\left(\frac{12}{27.73}\right)(25\,\text{ft}) - 280\,\frac{\text{lb}}{\text{ft}}(14\,\text{ft}) = 0$$

$$P = 362 \text{ lb/ft}$$

The correct answer is C.

Solution 2.16

No reference is required – basic construction drawing.

Option A is side view; Option B is top view; Option D is not related anyway.

The correct answer is C.

Solution 2.17

PE Civil Handbook: Chap 2 Construction (Section: Scheduling)

With an imposed finish date of 29 days, the forward pass and backward pass of the node diagram is shown below.

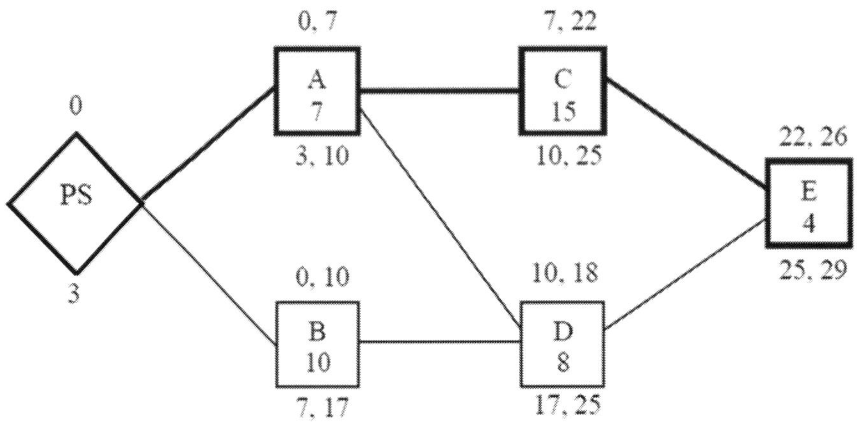

Activity	Duration	IPA	ES	EF	LS	LF	TF	FF
A	7	-	0	7	3	10	3	0
B	10	-	0	10	7	17	7	0
C	15	A	7	22	10	25	3	0
D	8	A, B	10	18	17	25	7	4
E	4	C, D	22	26	25	29	3	3

In the calculation of free-float (FF), we take the early-start (ES) and latest-start (LS) dates of the succeeding activities, whichever is smaller (earlier). In the case of imposed finish date greater than the calculated finish date, the difference appears as a free float only with the last activity(s) in the network, i.e., those with no successors.

The correct answer is A.

Solution 2.18

Part 1926, Subpart Z, Standard Number 1926.1103 applies to the 13 carcinogens. Hydrogen sulfide is not included in the list.

The correct answer is A.

Solution 2.19

PE Civil Handbook: Chap 2 Construction (Section: Scheduling)

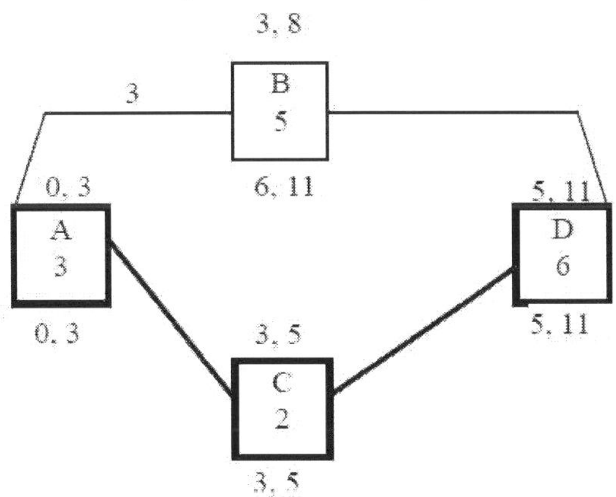

Activity	Duration (Days)	IPA	Type of Relationship	Lag	ES	EF	LS	LF	TF	FF
A	3	–			0	3	0	3	0	0
B	5	A	SS	3	3	8	6	11	3	3
C	2	A			3	5	3	5	0	0
D	6	B	FF		5	11	5	11	0	0
		C								

The correct answer is D.

Solution 2.20

PE Civil Handbook: Chap 2 (Section: Earthwork Construction and Layout)

Average End Area Method: $V = [(A_1 + A_2)/2] \times L$

Cut:

$V_1 = [(150 + 102)/2] \times (50 - 0) = 6{,}300\ ft^3$
$V_2 = [(102 + 110)/2] \times (100 - 50) = 5{,}300\ ft^3$
$V_3 = [(110 + 96)/2] \times (150 - 100) = 5{,}150\ ft^3$
$V_4 = [(96 + 81)/2] \times (200 - 150) = 4{,}425\ ft^3$

Total Cut = 21,175 ft^3 = 784.3 yd^3

Fill:

$V_5 = [(72 + 87)/2] \times (50 - 0) = 3{,}975\ ft^3$
$V_6 = [(87 + 105)/2] \times (100 - 50) = 4{,}800\ ft^3$
$V_7 = [(105 + 100)/2] \times (150 - 100) = 5{,}125\ ft^3$
$V_8 = [(100 + 127)/2] \times (200 - 150) = 5{,}675\ ft^3$

Total Fill = 19,575 ft^3 = 725.0 yd^3

The cut is larger than the fill. Therefore, surplus soil = 784.3 yd^3 – 725.0 yd^3 ≈ 60 yd^3

The correct answer is A.

Solution 2.21

PE Civil Handbook: Chap 2 (Section: Material Quality Control & Production)

$$\text{Absorbed moisture (\%)} = \left(\frac{SSD\ Mass - Oven\ Dry\ Mass}{Oven\ Dry\ Mass}\right) \times 100$$

$$= \left(\frac{2{,}354\,g - 2{,}165\,g}{2{,}165\,g}\right) \times 100 = 8.7\%$$

The correct answer is A.

Solution 2.22

PE Civil Handbook: Chap 2 Construction (Section: Scheduling)

Without formal analysis:

- Budgeted quantity of carpet installed in 4 days = (2,600/10 SF per day) x 4 day = 1,040. The subcontractor actually installed 1,500 SF of carpet. So, ahead of schedule.
- Actual cost in 4 days is $3,150, whereas the budgeted cost of work performed is 1500 SF x $2.30 per SF = $3,450. So, under-budget

The correct answer is A.

With formal analysis:

ACWP = $3,150
BCWP = 1500 SF x $2.30 per SF = $3,450
BCWS = (2,600 SF / 10 days) x 4 days x $2.30 = $2,392

CV = *BCWP* – *ACWP* = $3,450 – $3,150 = $300 (positive, under-budget)
SV = *BCWP* – *BCWS* = $3,450 – $2,392 = $1,058 (positive, ahead of schedule)

Solution 2.23

PE Civil Handbook: Chap 2 Construction (Section: Earthwork Construction and Layout)

Relative compaction = $\gamma_{d,field} / \gamma_{d,max}$ x 100
= 111.8 pcf / 118.5 pcf x 100
= 0.943 x 100
= 94.3 %
≈ 94 %

The correct answer is A.

Solution 2.24

The steepest slope of 1:5 is possible. The slope of 1 : 4 is still beyond the limit, as shown below.

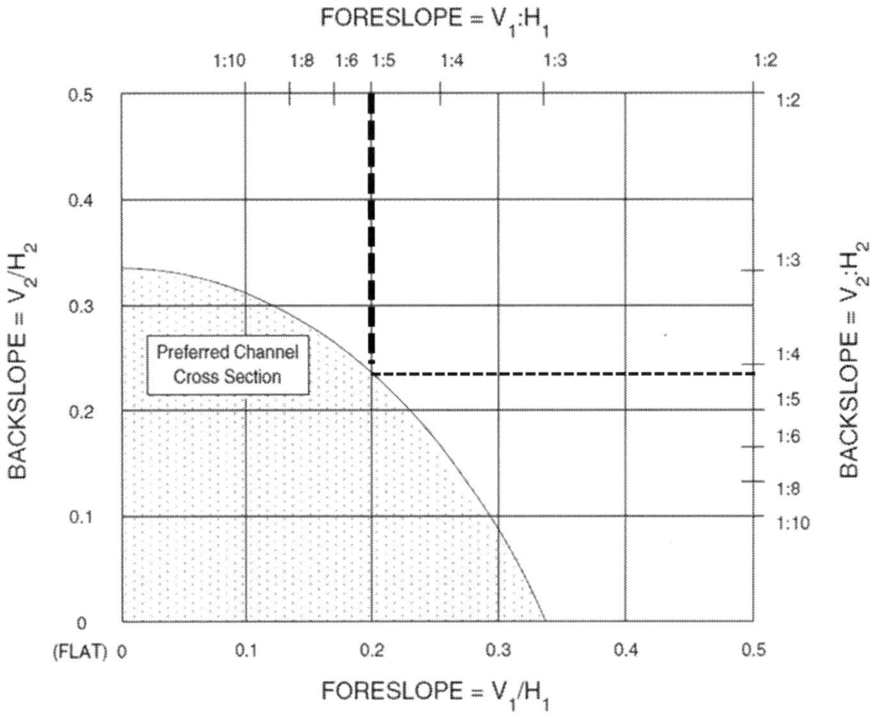

The correct answer is C.

Solution 2.25

MUTCD, 2009, Table 6C-4:

Table 6C-4. Formulas for Determining Taper Length

Speed (S)	Taper Length (L) in feet
40 mph or less	$L = \frac{WS^2}{60}$
45 mph or more	L = WS

W = width of offset = ½ (66 ft – 12 ft – 12 ft – 11 ft – 11 ft) = 10 ft
S = posted speed limit = 50 mph
Length of taper = $L = WS$ = (10 ft)(50 mph) = 500 ft

The correct answer is A.

Solution 2.26

Table states, the combined volume is 2/3 of the sum of the original bulk volumes. This means that the combined volume decreases to 2/3 of the original bulk volumes.
Sum of original bulk volume of the components = 0.20 m^3 x 3/2 = 0.30 m^3

Volume of Cement = 0.174 x 0.30 m^3 = 0.0522 m^3
Volume of wet fine aggregate = 0.391 x 0.30 m^3 = 0.1173 m^3
Volume of wet coarse aggregate = 0.348 x 0.30 m^3 = 0.1044 m^3
Volume of water = 0.087 x 0.30 m^3 = 0.0261 m^3

The correct answer is D.
[the new recent edition of EB001, *Design and Control of Concrete Mixtures*, 17/e, 2021, revised the given table, Table 13-16]

Solution 2.27

PE Civil Handbook: Chap 2 Construction (Section: Scheduling)

The correct answer is B.

Solution 2.28

No reference is required – basic calculation on construction equipment.

48 tons = 48 x 1,000 kg = 48,000 kg = 48,000 kg x 9.81 m/s^2 = 470,880 N = 470.9 kN
Force = Stress x Area
Area = Force / Stress = 470.9 kN / 200 kPa = 2.35 m^2

The correct answer is B.

Solution 2.29

PE Civil Handbook: Chap 2 Construction (Section: Earthwork Construction and Layout)
All the given options are true. The correct answers are A, B and C.

Solution 2.30

PE Civil Handbook: Chap 1 General Engineering (Section: Engineering Economics)

P = 155 (P/A, 6%, 8 yr) + 35 (P/G, 6%, 8 yr)
= \$155 (6.2098) + \$35 (19.8416)
= \$1,657
≈ \$1,660

Factor Table - i = 6.00%

n	P/F	P/A	P/G	F/P	F/A
1	0.9434	0.9434	0.0000	1.0600	1.0000
2	0.8900	1.8334	0.8900	1.1236	2.0600
3	0.8396	2.6730	2.5692	1.1910	3.1836
4	0.7921	3.4651	4.9455	1.2625	4.3746
5	0.7473	4.2124	7.9345	1.3382	5.6371
6	0.7050	4.9173	11.4594	1.4185	6.9753
7	0.6651	5.5824	15.4497	1.5036	8.3938
8	0.6274	6.2098	19.8416	1.5938	9.8975
9	0.5919	6.8017	24.5768	1.6895	11.4913
10	0.5584	7.3601	29.6023	1.7908	13.1808

The correct answer is B.

Solution 2.31

PE Civil Handbook: Chap 2 Construction (Section: Scheduling)

BCWS = \$95,000
ACWP = \$100,000
BCWP = Earned value = \$80,000

$CV = BCWP - ACWP$ (Earned – Actual) = \$80,000 – \$100,000 = –\$20,000
$SV = BCWP - BCWS$ (Earned – Planned) = \$80,000 – \$95,000 = –\$15,000
Both CV and SV are negative.

The correct answer is D.

Solution 2.32

No reference is required – basic calculation of construction operations.

Total fixtures = 24 + 2 x 36 = 96
Hours required = 96 x 1.5 hours each = 144
Number of days = 144 hour / 2 crew / 8 hour-day = 9
The correct answer is B.

Solution 2.33

70% is equivalent to 20 kg. So, 20% is equivalent to (20/70) x 20 = 5.71 kg

The correct answer is B.

Solution 2.34

PE Civil Handbook: Chap 2 Construction (Section: Construction Operations and Methods) The answer is not directly given in the Handbook but the concept is very basic for site preparation.

Total dozer cycle time is the sum of its fixed cycle time, doze time and return time. Fixed cycle time represents the time required to maneuver, change gears, start loading, and dump.

Dozing distance = 200 ft
From the Table, fixed time = 0.05 min
From the Table, dozing speed = 2.5 mph = 2.5 mph x (5,280 ft / 60 min) = 220 ft/min
Dozing time = Distance / Speed = 200 ft / 220 ft/min = 0.91 min

Return speed = 5 mph = 5 mph x (5,280 ft / 60 min) = 440 ft/min
Return time = 200 ft / 440 ft/min = 0.45 min
Total time = fixed time + doze time + return time = 0.05 + 0.91 + 0.45 = 1.41 min

The correct answer is C.

Solution 2.35

PE Civil Handbook: Chap 2 Construction (Section: Scheduling)

Activity	Duration	IPA	ES	EF	LS	LF	TF	FF
A	3	-	0	3	0	3	0	0
B	7	A	3	10	3	10	0	0
C	4	A	3	7	6	10	3	3
D	5	B	10	15	13	18	3	3
E	8	B, C	10	18	10	18	0	0

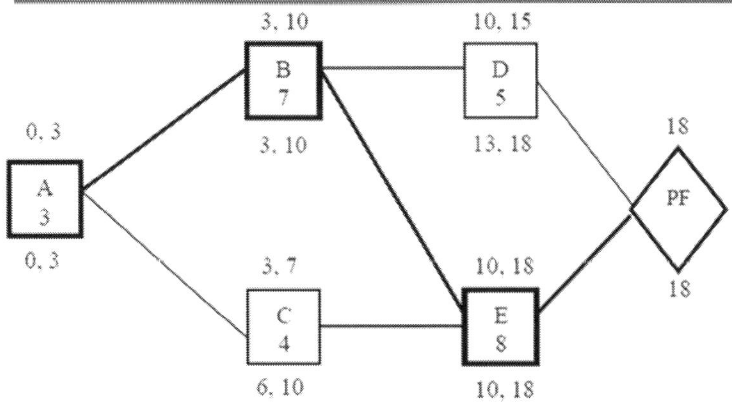

The correct answer is D.

Solution 2.36

No reference is required – it is part of tributary area and load calculation.

Point C will carry 50% of the girders' load from each side and the columns loads.
A W10 x 100, implies a wide-flange section with a nominal depth of 10 in. and a self-weight of 100 lb/ft.
Load from the column = 20 ft of W10 x 100 and 13 ft of W10 x 54
= 20 ft (100 lb/ft) + 13 ft (54 lb/ft)
= 2,702 lb

Load from the girders = left side + right side
= 16 ft of W18 x 40 and W18 x 50 + 17.5 ft of W21 x 44 and W21 x 57
=16 ft (40 lb/ft + 50 lb/ft) +17.5 ft (44 lb/ft + 57 lb/ft)
= 1,440 lb + 1,768 lb
= 3,208 lb

Total dead load at C = 2,702 lb + 3,208 lb = 5,910 lb = 5.91 kip
The correct answer is A.

Solution 2.37

PE Civil Handbook: Chap 1 General (Section: Mechanics of Materials)

P = 14 ton = 14 x 2,000 lb = 28 kip

$$\Delta = \frac{PL}{A_{ps}E_{ps}} = \frac{(28\,\text{kip})(310 \times 12\,\text{in.})}{(0.153\,\text{in.}^3)(27,500\,\text{ksi})} = 24.7\,\text{in.}$$

The correct answer is D.

Solution 2.38

PCA EB001, *Design and Control of Concrete Mixtures*, 17/e, 2021, pp. 109.

Durability cracking (D-cracking) is a series of closely spaced, crescent-shaped cracks near a joint, corner or crack, caused by freeze-thaw expansion of the large aggregates. See Fig. 5.23.

Cracks which form before concrete has fully hardened are known as the early-age cracks, or pre-hardening cracks. There are three main types of early-age cracks: Plastic shrinkage cracks, Plastic settlement cracks, and cracks caused by formwork movement.

If proper curing is not performed, due to the decrease in moisture, the concrete shrinks and cracks in block shapes which ca n be called shrinkage cracks due to improper curing.

The correct answer is D.

Solution 2.39

PE Civil Handbook: Chap 2 Construction (Section: Construction Operations and Methods)

Average speed, $S = 5$ mi/h
Compacted lift thickness, $L = 6$ in.
Effective roller width, $W = 10$ ft
Job efficiency = 0.75
Number of passes, $n = 4$

Compacted cubic yards per hour = $1/n$ (16.3 x W x S x L x efficiency)
= 1 / 4 (16.3 x 10 ft x 5 mi/h x 6 in. x 0.75) = 917
Note: the units are not consistent in this equation.

The correct answer is A.

Solution 2.40

PE Civil Handbook: Chap 2 Construction (Section: Earthwork Construction and Layout)

Swell factor = Loose dry unit weight / Bank dry unit weight =1,960 / 2,800 = 0.70

Loose volume = 1,000,000 LCY
Bank volume = 1,000,000 x 0.70 = 700,000 BCY

Shrinkage factor = Bank dry-unit weight / Compacted dry-unit weight = 2,800/3,500 = 0.80
Compacted volume = 700,000 x 0.80 = 560,000 CCY

The correct answer is D.

Some Useful Tips

- Always update yourself on the exam policy from the https://ncees.org/ website and from your friends who took the exam recently. The NCEES exam policy updates very frequently and much of the practical information is not available at the website.
- PE Civil Ref Handbook will be always open on one side of the computer screen. You can open one additional reference at a time.
- Try to memorize which equation is available at which reference.
- No calculation may be required for about 50% of the questions.
- The number of questions in the morning is not fixed to be 40. It may be slightly less or more. If you spend less time in the morning, you will get the remaining time back in the afternoon. More specifically, if you stop at 3 hr 30 min in the morning, your afternoon time will start from 3 hr 30 min.
- You may be able to access your Afternoon (Depth) reference manuals in the morning as well.
- Practicing the problems at home may take longer time as the home environment is not so comfortable.
- Learn to navigate to a particular topic using the Ctrl + F button of the computer keyboard.
- While practicing, use any NCEES recommended calculators. Carry 2 on the exam day.
- Don't forget to carry the photo ID on the exam day.
- If you fail the exam, you will lose the exam fees only - nothing else! Don't panic!
- Whenever you see a problem, first try to eliminate some options before start solving it.
- Whenever you see a problem difficult, skip it by flagging and move forward. Come back to the problem later. This strategy is very helpful to whip on the time.
- Some questions are long, however, the solutions are mostly short for long questions.
- You don't need to score 100% to pass the exam. About 70% of the score is believed to be good.
- The Breadth (Morning) exam may not be the same for everybody in the exam room.
- If you flag/skip a question you can get a chance to review them just before the final submission. You can go back to previous question at any time but cannot see the preview of all flagged/skipped questions until you go to the end of the exam.
- Save some time to get used to the computer based testing format.
- It is very difficult to concentrate on the last hour especially in the afternoon session. Carry some chocolate or Tylenol to keep your brain calm on the exam day.
- Do not forget to contact the author (islamunm@gmail.com) whenever you feel so.

Best wishes for your PE Civil Exam! Again, if you fail, you will lose your exam fees only – nothing else! Carry some chocolate and Tylenol and be confident!

Check out the author's practice exam for the PE Civil Breadth examination.